Amateur Radio Digital & Voice Emergency Communications

Build your community group's assets & expertise
2nd Edition – June 2017

Gordon L. Gibby KX4Z
BEE, MS, MD
and Barry Isbelle N2DB

ISBN: **1548004340**
ISBN-13: **978-1548004347**

VERSION 2.0
June 9, 2017

This manual is the property of

_____

who can be reached at:

_____

_____

COVER DESIGN
Emergency radio systems.
From Top to Bottom, Left to Right:
2-meter "go-box" based on Juentai JT-6188 / Signalink,
with built-in storage battery & charger. A second 2-meter
"go-box" based on Icom 28A, also with storage battery;
both include a blue "combiner" box that allows mic & digital
to work together; Middle 2-shelf homebrew box has an ICOM
automated long-wire tuner and a rolled-up HF antenna & houses an
HF Icom, as well as a Yaesu 2-meter, an SCS Pactor Modem;
on the tiles are multiple Baofeng hand-held transceivers,
a SURECOM digital simplex repeater controller and a
portable laptop as well as a roll of RG8X coax.

DEDICATION

This text, like the First Edition, is dedicated to my wonderful wife,
Nancy Gibby,
who has supported me through good times and sad times,
calmly puts up with my "projects"
on every available horizontal surface in our house,
and is one of the most faithful people I have ever known
to our Lord and Savior.

PREFACE TO THE SECOND EDITION

This book is the result of ongoing learning and development by those of us in the Alachua County Florida ARES (Amateur Radio Emergency Service) group.

In the first edition, I explained relatively inexpensive ways to build digital emergency communications gear and networks, based largely on free `linbpq` software by John Wiseman G8BPQ, and additional software by Andrei Kopanchuk UZ7HO and John Langner WB2OSZ.

At the time I wrote the first edition, our group had created a number of Raspberry Pi-based digital nodes, repaired a crucial "high-perch" node, and were gaining skill with important techniques as WINLINK.

Since that writing, our group has increased the pace of our learning, realizing (belatedly) the incredible power of UZ7HO's EasyTerm software to provide local emergency digital communications and now even dipping our toes into MESH communications.

Thanks to the amazing helpfulness of the Florida Forest Service, we were able to put up an additional "high-perch" node, and now have far better emergency digital coverage of the western side of our county --- and well into several counties west of us.

We tackled an incredible Hurricane-based Full Scale NIMS-compliant Exercise and actually succeeded for the most part. We had more participation, more messages transacted, in a more difficult and constantly changing exercise, than I believe has ever happened in our county before.

In this second edition, I try to cover those important new tools for effective emergency digital amateur radio communications. As the ARRL has pointed out: digital communications are going to be the *sine qua non* for emergency communications in the future.

Gordon L. Gibby MD KX4Z
Newberry, Florida

Software Designation Note

In the text, the same application may be called by several similar names – for example, consider UZ7HO's excellent software application that exists as "term.exe" on Windows computers, but is downloaded from a link that currently reads "easyterm39.exe" (previously smaller numbers for earlier versions) and installs itself in a directory "easyterm39." In the text, it is variously referred to as EASYTERM, EasyTerm, easyterm39, and easyterm39 (term.exe). Soundmodem likewise is denoted by various names: Soundmodem, soundmodem.exe, etc.

CONTENTS

ACKNOWLEDGMENTS

I would like to acknowledge the incredibly patient members of the Alachua County ARES group, who have made this entire project possible. Hours and hours of fun times together learning how to solder, hurl weighted fishing lines over high branches, decipher cryptic computer screen, build odd-looking antennas, and connect up all manner of electronic equipment to make miracles of amateur radio wireless communication become the norm in North Central Florida.

Jeff C.
Tom C.
Rosemary J.
Jim B.
Art & Cindy
Harry M.
Susan H.
John T.
Chris C.
Cheryl C
Allan W.
Larry R..

And a special note of thanks to Jim B., who proofed the entire first edition text and the changes for the 2nd, along with Jeff C.! Additional thanks to Dave Welker W2SRP whose information allowed the chapter on Marion County Hospital Emergency Communications, and to Barry Isbelle, who authored a chapter giving insights into how SEDAN as well as WINLINK are used in Orlando, Florida.

1 COMMON OPERATING PICTURE AND SITUATIONAL AWARENESS WHEN NORMAL COMMUNICATIONS FAIL

Common Operating Picture (COP) and **Situational Awareness** (SA) -- those are rock-bottom requirements to enable effective disaster coordination, decision-making and incident response. In a disaster severe enough to destroy or overwhelm normal communications, the ability for critical leadership to even *know what is going on* can be completely lost. **It is the goal of this book to assist in developing backup amateur radio communications that will make it possible to keep that Common Operating Picture visible, even when normal commercial communications have failed. And then, to be able to communicate both locally and to distant resources, to request, obtain and direct assets as needed.**

"**Common Operating Picture:** A continuously updated overview of an incident compiled throughout an incident's life cycle from data shared between integrated systems for communication, information management, and intelligence and information sharing. The common operating picture allows incident managers at all levels to make effective, consistent, and timely decisions. The common operating picture also helps ensure consistency at all levels of incident management across jurisdictions, as well as between various governmental jurisdictions and private-sector and nongovernmental entities that are engaged. "

"**Situational Awareness:** The ability to identify, process, and comprehend the critical elements of information about an incident"[1]

Without enough information to form a common operating picture among the decision makers, including the Incident Commander and the EOC (Emergency Operations Center), it will be very difficult for them to respond wisely or appropriately to the problems. Cell phones, trunked radio systems and other communication systems may still function in an emergency. Many emergencies don't destroy all normal communication. Even then, they may be overwhelmed by the increase in the flow of information, but at least there *is* information.

The real problem comes when the disaster **destroys or significantly damages normal communications**. Such was the case during Hurricane Katrina and the resultant flooding along the Gulf Coast, and it was prominently and frequently mentioned as a major cause of the flawed response, in the subsequent Congressional review. Even when the planned new **FirstNet next-generation communications system** becomes a (very expensive) reality, there will be a need for backup communications --- FirstNet's communication bandwidth of a few tens of MHz in the 700 MHz band is not at all large compared to amateur radio allocations, and FirstNet may well depend on exactly the same cell phone towers whose destruction by the incident took down commercial communications.[2]

VOICE OFTEN SUFFICES

When normal communications are overwhelmed or disrupted, yet the disaster is small, the amount of information that must be transferred --- *where the problem is, what kind of incident occurred, what risks remain, which roads are blocked, which power lines are down* -- is not enormous, and the transfer can frequently be aided by **voice** amateur radio communications. A well-functioning EOC will have some computerized system to log the incoming data points, and create an overview that allows the Common Operating Picture for everyone with access to the web-based or other data synthesis software (e.g., WebEOC or EM Constellation). Amateur radio **voice** communications, often by VHF/UHF FM, often using repeaters, can transmit information easily by modestly-trained volunteers.

BUT VOICE CAN BE INADEQUATE....

However, when the disaster is large, and the communications losses are severe, the need for **data** dramatically increases. In Katrina, there were literally *thousands* of requests for help, *thousands* of resources to be tracked, directed, and managed. It is virtually impossible for this level of information can be managed accurately and efficiently by voice communications alone. It is this level of need for which emergency amateur radio services need to be be prepared, and trained.

The disaster that requires huge data transfers is the one where the community will be most desperate.

It is a tragic mistake to fail to train adequately for the exact scenario in which your resources will be MOST NEEDED. Yet that is what many well-meaning volunteer groups do. The example of a modern army is insightful: yes, they train for helping with food and water delivery during storms, but what they put the greatest training into, is preparing for all-out WAR.

Normal daily business practices also give great insight here. Voice communications suffice for trivial information transfer, but most businesses heavily depend on a <u>fax machine</u> at the least, and much more likely, <u>email</u> or even <u>dedicated customer order tracking software</u>. Beyond that, a successful business probably has materials management systems, process flow software, and of course, financial software. The amount of data that moves in and out of a modern corporation is *huge*. And if there is a real disaster, one that disrupts the Internet, and possibly even satellite communications, that data comes to a halt.

Without data flow, a business can't see incoming orders, and an EOC can't see the thousands of urgent needs. Without data flow, a business can't keep track of the progress of ordered raw materials, and an EOC cannot order, track, and direct the delivery of thousands of truckloads of special equipment, food rations, water, and medical supplies. Without data flow, personnel in a business can't be directed, and the same is true of an EOC.

During Katrina, despite days of advance warning,

- 37 of 41 broadcast radio stations in and around New Orleans were knocked off the air.
- Thirty-eight 911 call centers went down *So much for capturing citizen cries for help!*
- 20 million phone calls did not go through on the day after the hurricane.

- More than 3 million telephone lines went down. *Can't even call for help -- or direct aid workers*
- The police headquarters of New Orleans, as well as 6 of 8 police district buildings, were out of commission.
- New Orleans police communications system failed and remained failed for three days. Only 2 channels of a backup system remained for **hundreds** of emergency personnel who were already overwhelmed.
- Louisiana state police communications were dependent on 46 towers to support 10,000 users. After loss of electrical distribution, tower after tower ran out of power when their stored generator fuel was exhausted. Refueling was nearly impossible due to flood waters and debris. [3]

The Mississippi National Guard was reduced to using runners to ferry messages from Commander to Commander. As one spokesperson put it, "In other words, we're going to the sound of gunfire, as we used to say in the Revolutionary War."[4]

With that level of destruction of communications, you can begin to imagine the level of confusion and loss of the big picture---loss of command and control -- that occurred as a result!

DIGITAL VASTLY INCREASES THROUGHPUT

In a massive disaster vast numbers of messages must get through, in order to maintain common operating picture, situational awareness and effective command and control. Even with pedestrian amateur radio digital systems, such as Packet (AX.25), MT63 and others, the throughput of digital communications far outpaces what can be reliably and faithfully delivered by voice communications. Essentially, voice communications are not only hampered by normal static and poor signals in communications systems, but also **require data entry at the receiving end**, in the form of either handwriting or typewriting to create a written version of the data that can be promulgated to others. Normal handwriting can't keep up 20 words per minute (wpm) for long. Most typists are in the range of 40-70 words per minute -- and it is exceedingly rare that this level of throughput can be maintained using voice channels.

By comparison, digital systems have inherently **far higher throughput**. Digital systems come in at least two flavors: "broadcast" (non-ARQ) and "error-corrected" (ARQ -- "Acknowledge-Request"). (Broadcast signals can be further broken into those with forward error correction and those without, but neither can be guaranteed error-free.) As the name suggests, broadcast systems have no inherent security! Error corrected systems typically require a form of acknowledgment / non-acknowledgment and handshaking to allow repeat transmission of frames ("packets") of data that were found to have errors in the transmission. Typically such error-corrected transmissions are a one-to-one transmission, whereas non-error-corrected modalities allow for broadcast from one-to-many, but without an ironclad guarantee of perfect transmission.

Table 1-1 compares an estimate of voice transcribed throughput to one relatively high speed broadcast-type (non-error-corrected) radio digital mode, and the error-corrected modes utilized by the proven WINLINK email amateur radio transmission system. Figure 1-1 shows graphically the same data.

TABLE 1-1 COMPARISON THROUGHPUT, VOICE VERSUS BROADCAST OR ERROR CORRECTED AMATEUR RADIO

MODE:	VOICE	MT63-2K ("Broadcast")	WINLINK Error-corrected
Most apparent limiting factor	writing/typing speed of receiving station		
Peak sustained throughput, words per min (word=5char)	30 wpm (estimate sustained accurate typist radio copy speed)	200 wpm [5] My testing: 240 wpm	PACTOR P3 2,400 wpm estimate [6] WINMOR > 1000 wpm (actual measurement of large file)
Estimated 50-word messages per minute	0.30	2.0	WINMOR - measured 3.02[i]

---

i   Actual test results carried out with optimal signal strength, twenty 250-character (50 "word") individual email messages

Advantage over VOICE	1.0	566% faster	906% faster (for large files 6,800% faster)[ii]

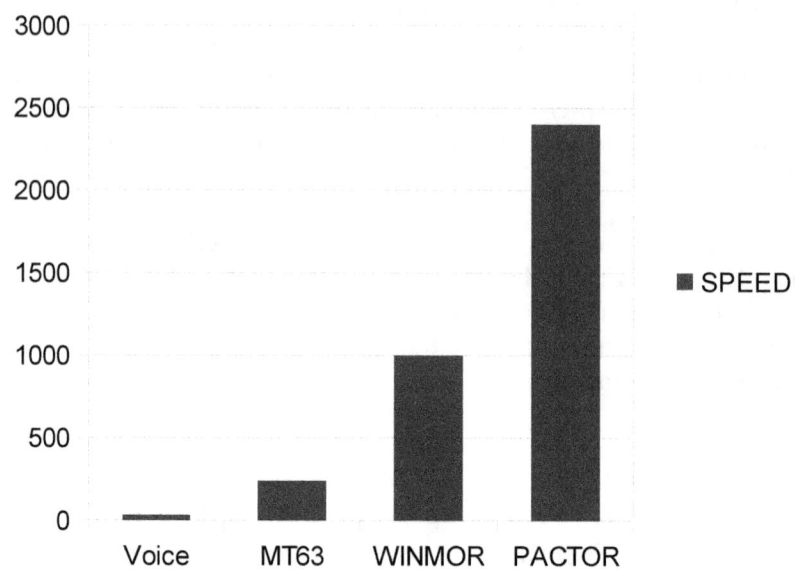

Figure 1-1. *Speed in word per minute of voice versus several digital modes, for "record" communications (which must be written down accurately). Voice compares rather poorly, because it is basically limited by a 30 wpm typical typing speed copying actual radio comms.*

TIME COURSE

An Emergency Operations Center has been described as a "coordination" center. In many ways, their mission is to support people on the front lines, and coordinate the orderly movement of resources towards the actual needs.

Communities themselves generally include their own life-sustaining businesses and services: food distribution and marketing businesses; hospitals and clinics; fuel distribution and marketing establishments;

---

ii Actual test results carried out with optimal signal strength; 30,000 word file for "large file" vehicle and machinery repair services, and so on. During the acute phase of a widespread disaster, these community resources may be temporarily shut down or hampered as citizens take shelter at home. Essential services that have to keep going (if physically possible) include hospitals, ambulances, police, and fire services. The EOC coordinates in an effort to assist these services in the acute phase. If normal communications fail, amateur radio can step in and help to keep information flowing.

After the acute phase is over, hopefully national resources will begin arriving within 3-5 days.. Amateur radio can't hope to provide ALL the communications needed to keep an entire community going, even if there were not the restriction against carrying on "business" via amateur radio. The sheer volume of data needed to keep all of the transactions flowing (likely Gbytes to Terabytes / day) in a community would vastly overwhelm the relatively tiny throughput of amateur radio. This is where commercial providers must step in and re-establish high speed wired, wireless and satellite communications.

Unfortunately, in truly large disasters, the "cavalry" commercial communications providers may not arrive in the hoped-for 72 hour period. Amateur radio operators may be providing badly needed communications in the worst disasters for much more than the mythical 72 hours.

Amateur radio volunteers who want to be ready to serve their communities in time of real need, should prepare not only for voice communications, but also for digital data modes, including broadcast type modes as well as error corrected modes.

In the next chapter we'll examine in more detail what are the most likely communications modes that can be leveraged by amateur radio in emergency communications to meet those needs.

2 FOUR KEY COMMUNICATIONS MODES ABILITIES FOR EMERGENCY COMMUNICATIONS

There are four basic communications modes I think will be of primary importance in emergency amateur radio communications. Competence and flexibility are the key traits of importance; you need to have more than one skill in your bag of tricks, and you need to be able to move back and forth between different communications technologies in order to meet the changing needs.

This chapter will not so much go into the technical details (that's for later in the book) as much as it will give you the 30,000 foot view of WHAT you can do, if you have both the assets and the skills.

1. VOICE: This is usually the easiest, and most common--but important! You must be able to send and receive voice communications. Vital information can be rapidly discussed over voice communications easily. Within a community, voice transmissions are easily carried out over VHF (2 meter) or UHF (70 cm and other bands) FM transmissions. Voice repeaters are often present, dramatically increasing the range. An emergency communications group should both harden those repeaters against threats (including CME/EMP) and further, provide for communications alternatives should the repeaters fail. That might include emergency emplacement of backup voice duplex repeaters, or of store-and-forward simplex replacement repeaters positioned on the tops of (remaining) high buildings, as well as adequate simplex communications equipment. (See Chapter 12.)

- - - - - - - -

*All of the remaining important emergency communications modes involve **digital communications,** so here's* an overview of how digital compares to voice.

- - - - - - - -

30,000 Foot Overview of Digital Communications

Digital isn't hard or obscure: there are several available hardware/software packages that allow for multiple different modes of digital communication using normal amateur transceivers (FM or SSB, single-sideband). Downloads of popular free software packages already number into the *hundreds of thousands*. Sales of digital connection packages are so brisk that there are frequent backlogs!

The key is to recognize that just as with voice transmission, information is usually going to be fed (as some form of *audio*) into the transceiver microphone input, and information is going to be extracted from the receiver audio output. (With more advanced digital systems, the audio stages may be bypassed for direct connections to modulating/demodulating stages.) The required hardware systems to accomplish this have gone through some evolution as hams worked out progressively easier ways to make digital work.

In the beginning days of amateur radio digital communications, a specialized hardware device known as a Terminal Node Controller (TNC) performed the interface between
a) mic input/speaker output and
b) the terminal—now a computer--used to see and create the digital information.

These hardware TNC devices (including the ubiquitous Kantronics KPC-3) are still commonplace and work well for many types of digital modes, including Packet (AX.25) and even PACTOR.

Rapidly advancing personal computer power brought about a big change. The ready availability of powerful **computer soundcards** has caused a pleasant revolution wherein the amateur radio transceiver is connected to mic and speaker connections of the soundcard, and then software performs the functions previously requiring dedicated TNC hardware. The **Tigertronics Signalink-USB** is a popular external soundcard-based interface device that provides a high quality soundcard as well as transformer-based isolation between the amateur radio transceiver and the computer. So many of these have been sold that there are now several competitors – including the earlier Rascal devices, the Timewave Navigator, and recently the MFJ-1204, and others. This book will also discuss a homebrew kit, one of many do-it-yourself examples of ways to build similar interfaces yourself. (It turns out that the MFJ-1204 design[7] is actually very similar to the homebrew soundcard interface described later in this book.)

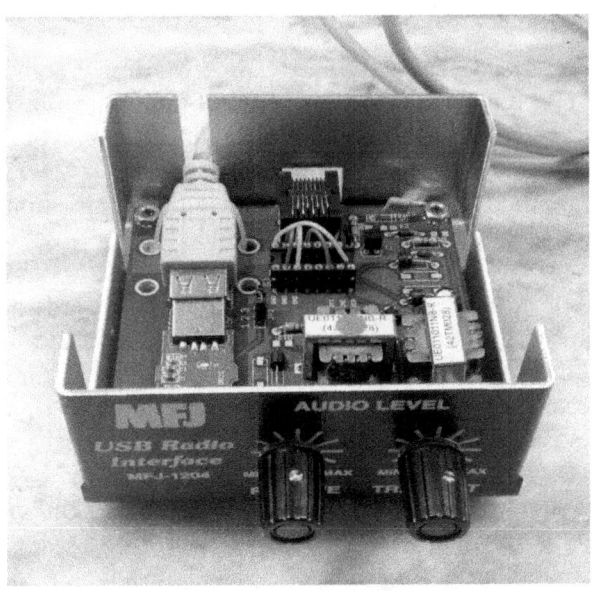

Fig. 2-1 *MFJ 1204 USB Radio Interface, a competitor to the popular Signalink USB. Multiple different accessory cables are available to connect to different radios. Very similar to homebrew soundcard design (when connected to a USB dongle soundcard) included later in this book. The soundcard portion of the MFJ device is visible at the left, with a USB cable permanently plugged into it.*

Multiple, and often free, software packages exist to function with these soundcard-based solutions (e.g., FLDIGI by W1HKJ, easyterm39 (term.exe) and soundmodem.exe by UZ7HO, older AGWPE and others), to allow for a dizzying array of digital modes to be transmitted and received. A partial list of digital modes now includes:

CW (the "original digital" can be done by computer)
PSK
Olivia
Thor
Contestia
MT63
RTTY

Most of these modes include many sub-modes of different speeds and resultant bandwidth. Generally, wider bandwidth and increased throughput go hand in hand.

An extremely popular and free software for digital communications primarily on HF, but also on VHF is FLDIGI (and related packages).[iii] A popular (and free) software for AX.25 keyboard-to-keyboard and file-transfer packet communications is EASYTERM by UZ7HO. (Both of these have important emergency communications applications as will be discussed.)

With all these systems/software, ultimately keyboard characters or computer file contents are converted into audio signals and fed into the microphone input of the transceiver. The audio output from the transceiver goes into the soundcard & software converts it into digital signals that eventually show up as words on a computer screen, or at a higher level, email in an inbox --- all able to be delivered by ham radio alone in a disaster, with no need for cell or landline, or Internet.

This introduction having been completed, let's look into the 2nd, 3rd and 4th important ham radio technologies for ham radio communications.

2. WINLINK EMAIL: Although current amateur radio transmission systems (other than microwave MESH systems, discussed later in this text) aren't always fast enough to allow for real-time full-blown tcp/ip internet-type data communications (and indeed, due to the proscription of commercial traffic, it is probably inadvisable) they perform admirably well for email messages and even attachments of very limited size (e.g, < 20Kbytes). The WINLINK organization has decades of development and service in routine seagoing and emergency email routing and delivery over both HF and VHF/UHF amateur radio, and is becoming a defacto standard for emergency communications. For some groups, this is their primary – possibly ONLY --- emergency communications technique.

---

iii The "FL" series of free software not only includes keyboard-to-keyboard FLDIGI, but builds upon that with additional layers that offer error-correction and error-free message transfer.

WINLINK has the ability, using HF (shortwave) simple digital ham radio stations, get email in and out of a devastated area by connecting to volunteer gateway stations anywhere else in the nation or even the world. Other than satellite communications, this is the *only* system that is virtually independent of nearby ground-based infrastructure.

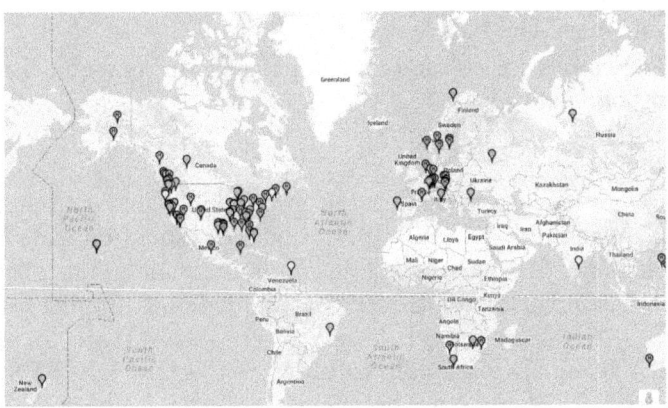

Figure2-2: *WINLINK HF (Pactor) Radio Message Servers world-wide.*[8] (2016)

The WINLINK organization has volunteer HF radio message station (RMS) servers scattered all around the globe. There are more than fifty on HF frequencies in the United States and two or three times that number on VHF frequencies (usually employing AX.25 Packet communications).

Winlink volunteers have served in an incredible number of situations and disasters. A partial list includes:

- Hurricanes Katrina, Rita, Wilma, Dolly, Dennis, Jeanne, Ivan, Frances, Charley, Isabel, Sandy
- Slovenian Weather Disaster
- HMS Bounty Rescue
- Haitian Earthquake Disaster
- The Asian Tsunami
- Western US Flood and Fire Relief
- Tennessee Tornado Outbreak 2008

- North Carolina Agency Fiber Optic Cable Failure
- Failure of IntelSat 804
- Indian Coastal Weather Disaster
- Chilean/Peruvian Weather Disaster
- Assisting the US Coast Guard; Locating lost and overdue vessels
- Australian Outback Communications
- Assisting NOAA National Weather Service, and their MAROB Program
- Gulf War "The Last Voice from Kuwait"
- Connecting doctors and remote patients during medical missions, and often at sea[9]

Your emergency communications group will want, at a minimum, to gain expertise at client-side WINLINK email transmission and reception, over both HF and VHF, giving you the ability to maintain local and distant email/attachment connections despite regional Internet failure.

WINLINK email-over-radio software (WINLINK EXPRESS) looks and works much like any other basic email software in the creation, reading, replying, and forwarding of email.

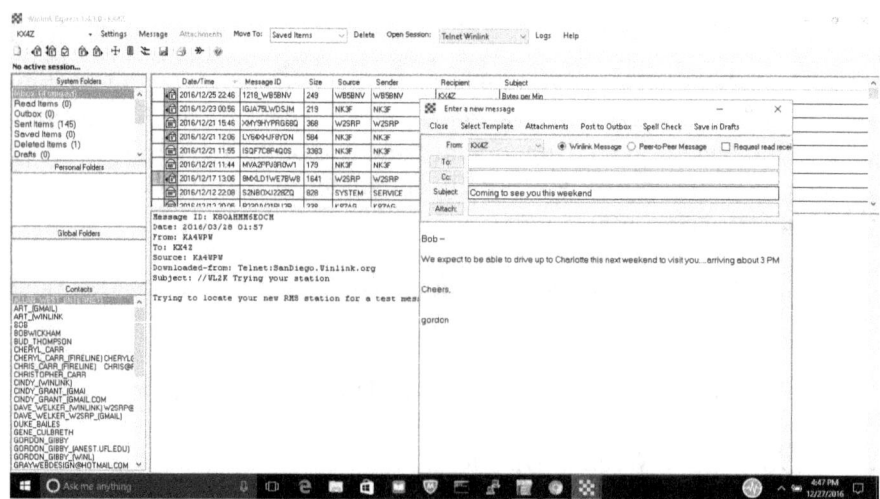

Figure 2-3. *WINLINK EXPRESS email application used to create a new email.*

Where it differs is in the transmission and reception. With normal

Internet email packages you merely hit "SEND." With WINLINK you may have three or four possibilities for transmission modalities! (E.g., high frequency PACTOR, high frequency soundcard-based WINMOR; TELNET over the Internet, AX.25 PACKET over VHF, and possibly even MESH over microwave.) As a result, there is an intermediate step of "posting to the OutBox" and then a final step of picking your connection method, establishing a connection, and allowing transmission/reception of email.

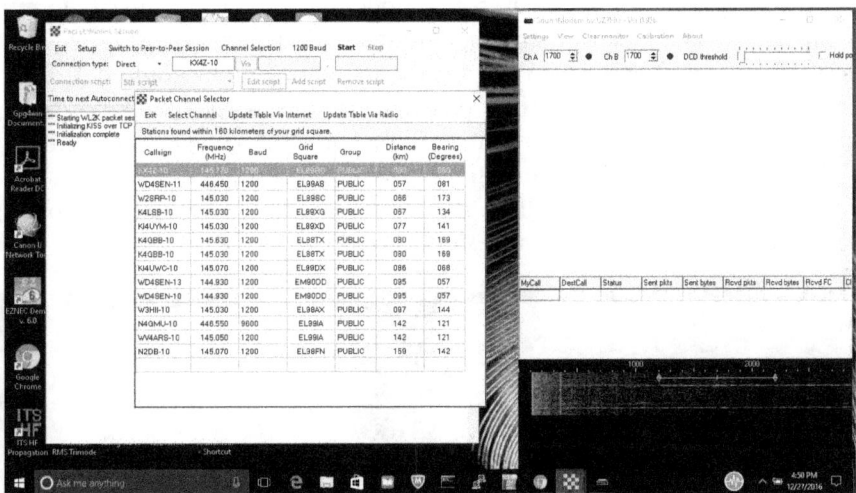

Fig 2-4. *WINLINK Packet Connection to transfer email. Soundcard software UZ7HO-soundmodem.exe on the right hand side, with a black-and-white FFT (Fast Fourier Transform) "waterfall" in the lower portion of the screen. On the left, WINLINK listing possible 2 meter packet servers for connection.*

ADVANCED EMAIL SERVICES

As your group gains expertise, you will gain the ability to funnel email from non-amateur emergency officials, and possibly even limited numbers of the general public, via special WINLINK capabilities[iv] [10] that piggyback those emails on the coattails of a licensed amateur control operator. Additionally, your group will want eventually to gain systems

---

iv Windows-based PACLINK software from WINLINK has this capability, as well as a Raspberry Pi-based Unix system available for download from a 3rd party.

operator ("sysop") abilities to run a WINLINK server, known as a Radio Message Server (RMS).   None of this is difficult, but all expertise requires time to develop.

3. PACKET CONTACT: Perhaps the biggest discovery our group made in emergency communications since the publication of the first version of this book, was a modern, Windows-based packet communications software, EASYTERM, created and offered free of charge by UZ7HO. It makes use of inexpensive sound-card technology by way of companion software soundmodem.exe.   Due primarily to my inexperience and lack of mentors, at first we didn't realize just how powerful this software and VHF packet communications could be!   Not only does it make packet CHAT possible (see below) but even more importantly, it allows for very quick, direct contact between one digital VHF station and another.   It reminds me of the current digital system employed on marine vhf radios, that allows you to "hello-up" any boat, so that communications can be established.

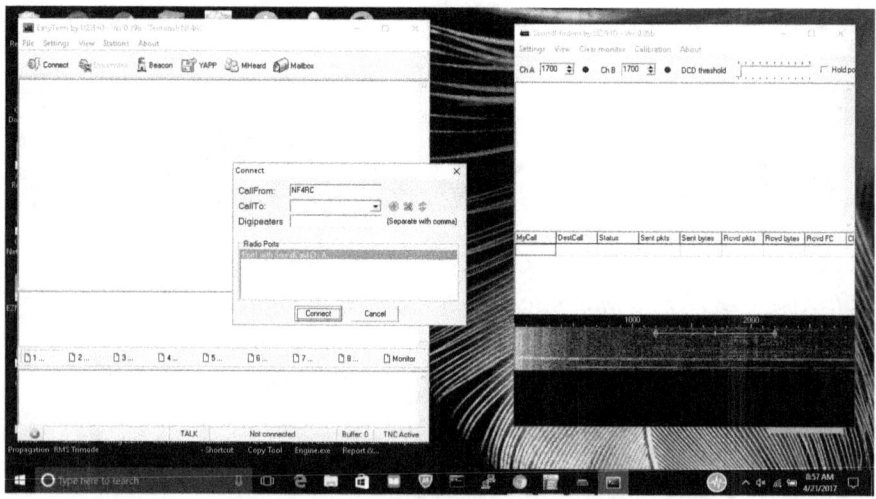

Fig. 2-5 *EASYTERM (left portion of screen) provides a simple Windows-based user interface to do classic keyboard packet communications easily.   Type into one pane, hit ENTER to transmit, read responses in another.   Connects to a soundcard using AGWPE interface, provided by (among others) soundmodem.exe (shown in right portion of screen).*

With packet communications, and an understanding of the available digipeaters and nodes in your area, you can sequentially step your way through an amazing number of intermediate digital relay stations and make contact with any digital station that has possible radio contact with any connected digipeater/node. You can then begin putting characters on their screen, sending tactical and other bulletins letter-perfect right to their computer. Cut/paste/copy standard techniques allow you to do this with a minimum of typing.

But it doesn't just stop there -- you can use the YAPP protocol built into EASYTERM to send an error-free digital file right to their computer --- without any intervention needed on their part! Nothing tricky at all. This makes it a cinch to transfer disaster-related spreadsheets, small databases, lists -- the kinds of stuff that are crucial to the administration of a disaster in an orderly fashion. It is no wonder that so many emergency groups across the United States still make very significant usage of VHF packet communication. EASYTERM makes this accessible for Windows users with soundcard systems, just as previous software made it possible on earlier operating systems.

Distance Advantages

This kind of user-controlled VHF digital communication requires more equipment and more user training than simple FM voice --- but it has a huge advantage over VHF voice for emergency communications: without any need for the Internet, the user can reach great distances to make urgent contacts, controlling their own repeating through nodes and digipeaters as necessary. Much father distances than typical voice repeaters can reach. As a concrete example, the SEDAN system in Florida allows for emergency communications across hundreds of miles.

4. "INLINE DIGITAL": This book will explain how it is possible to have the ability to seamlessly and effortlessly move from voice communications to sending digital tones/signals over the same frequency. For a concrete example, think about detailed weather bulletin information being disseminated from an EOC to multiple shelters during a hurricane. The information has many details and may discuss multiple nearby cities and towns from which shelter residents may have fled. Rather than drone this detailed information across a voice channel for many minutes, it can be sent as a broadcast type digital bulletin (e.g., using MT63-2000) which is copied simultaneously at multiple shelters, by ham radios and attached computers. At each shelter, the bulletin can then be disseminated in many different ways --- it can be printed out, handed out, read aloud, or even projected on a screen where interested residents can take in the news at their own speed.

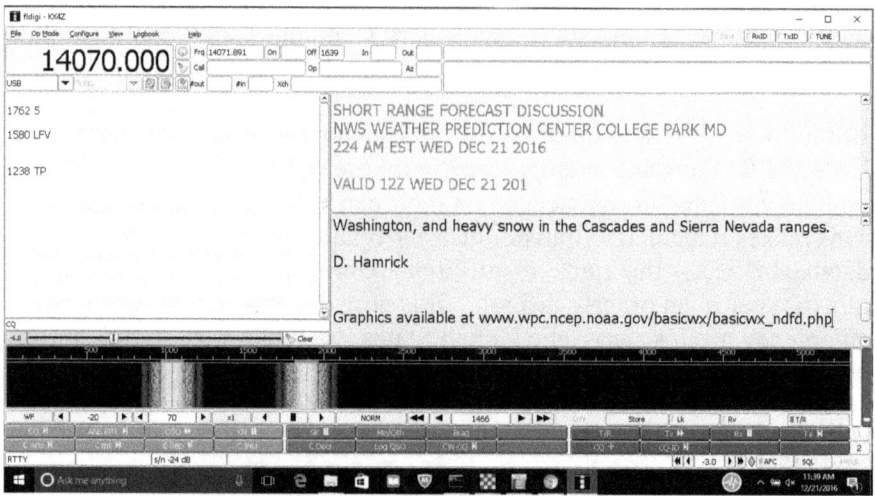

Figure 2-6. FLDIGI or similar software can be used with an appropriately configured transceiver to allow seamless switching back and forth between receiving (or sending) weather bulletins and other detailed information, and normal voice communications.

Once the bulletin is disseminated via digital text, voice communications can be utilized to answer specific questions if there are any.

I've listed this technique last in the list, because you can accomplish similar dissemination of information with other methods, such as individual packet connections, WINLINK email to multiple recipients, and even packet CHAT, as discussed later in this text. Those alternatives aren't quite as efficient as simple broadcast techniques with FLDIGI, but groups may choose to limit their training to prevent "learning fatigue" in their members ("Keep It Simple....."). WINLINK and keyboard packet connections are likely higher priority.

ADDENDUM

Packet CHAT

VHF AX.25 packet CHAT facilities offer a useful alternative "broadcast" way to have much the same effect as "inline digital." It isn't difficult to have two VHF systems operating on frequencies a MHz or so apart --- one can be running voice while the other is running what I could call "classic packet" -- giving the user the ability to check into a CHAT facility provided by many digital NODES, including KA-nodes and BPQ (or linbpq) nodes.

Although it isn't FAST over 1200 baud packet, everyone can see what anyone types. Its a huge "party line" where emergency shelters can all get the same information simultaneously from information sources logged into the CHAT system. And they can ask questions and get them answered, so that everyone see the answers as well. (Private remarks are also supported by some versions.)

CHAT tends to be a line-at-a-time affair so large bulletins' formatting would be affected, but I'm told it is used very effectively by emergency shelter-based amateurs in Orlando, Florida. UZ7HO's free software for Windows sound-card based systems, including easyterm38 / soundmodem97[11], brings these folks right up there with those using legacy hardware TNC's in their ability to do basic packet operations such as this. Over high speed MESH networks, CHAT or an equivalent web-based roundtable or even VOIP conference phone would be very effective.

TECHNICAL NOTE:
Soundmodem.exe provides for two different computer port interface protocols -- the older AGWPE interface, as well as the KISS interface. EASYTERM utilizes AGWPE. WINLINK EXPRESS software (see below) utilizes KISS. While soundmodem.exe can provide both interfaces, you can't *use* both simultaneously on one radio!

Gordon Gibby KX4Z

3   BUILDING TEAM COMPETENCIES AND ASSETS IN NORTH CENTRAL FLORIDA

NOTE: *To protect the privacy of my friends & colleagues, I've used pseudonyms for those who didn't give permission for their names to be used.*

To help others develop the kinds of emergency communications skills that were outlined in Chapter 2, I'll give a bit of the history of how things have progressed in North Central Florida -- home to a strong and thriving ham radio club, multiple voice repeaters, wonderful amateur radio operators, and *virtually no digital skills* when I began to be involved in emergency communications.   I may get a few details out of sequence, but we all had a great deal of fun learning all this stuff and developing all the solutions that will be detailed in the following chapters.

We all have strengths and we all have weaknesses. Leadership might be the gift to maximize others' strengths and cover their weaknesses, while molding people together toward a common goal and being patient enough to let it mature.   Leadership might also be serving others to help them grow.

The Start
    It started with WINLINK.[12] I became involved in WINLINK when I recognized the need for emergency communications after various dire possibilities, such as the cyber-takedown of the electrical grid written about in Lights Out by popular news anchorman, Ted Koppel.[13]   The

U.S. FEMA and Department of Homeland Security capabilities have indeed grown considerably since 9/11 and Hurricane Katrina, but Koppel and others have warned of multiple different persistent and significant threats to our culture. More and more people are becoming aware of the significant EMP risks from rogue nations and terrorist groups. WINLINK sysop experience serving mariners on the high seas solidified some digital capabilities in my repertoire that had begun with some fun experience with a homemade soundcard interface to a vacuum tube Heathkit transceivers using FLDIGI communication modes..

Multiple FEMA courses are online and available for free.[14] **Almost any involvement in emergency communications requires courses 100, 200 and 700**. (See Chapter 4 for more details.) Completing those gave me a background in the terminology. I also found that the local hams had relatively little experience with the full-blown introductory ARRL EC-001 emergency course, so I signed up and gained that experience -- which takes quite a few hours but is worth the time.[15] Here and in Chapter 4, I enthusiastically recommend all these courses to anyone who wants to try and develop a stronger emergency communications preparedness in their local community.

> Take the time to develop your own emergency communications credibility by investing time in freely available courses that give you written proof of completion.

First I began to be involved in local ARES group meetings, and took stock of the strengths of weaknesses of their setups at the local EOC and Red Cross.

INITIAL STRENGTHS & WEAKNESSES
- The local Red Cross had capable VHF/UHF voice radios and antennas.
- The Red Cross had an older solid state HF transceiver, and no antenna for it..... at all.
- The EOC had a very capable HF computer-controllable solid state HF transceiver, but a nearly worthless HF antenna.
- The EOC had two capable VHF/UHF transceivers, and modestly high (10-15 feet) commercially-available antennas.
- Neither facility had ANY digital capability.
- The EOC had backup power; the Red Cross did not.

- Neither facility appeared to have hardened their ham gear against EMP (Electromagnetic Pulse), though the EOC was probably proof against CME (Coronal Mass Ejection).

The local ARES group was at that time focused on providing emergency communications to up to a dozen hurricane shelters, primarily using handheld VHF transceivers. There was about a 50/50 breakdown between Technicians and higher-class license holders, and almost zero digital experience. The local club was focused to some extent on voice repeaters.

There was ONE packet digital node station in our community[16], but thankfully with a commanding physical position and strong signal. My knowledge of packet at that time was quite limited. My WINLINK HF station was the only RMS capable of moving email in or out of the city.

The group had two notable strong points:
a) several of the ARES members had actual experience--years of it -- in emergency communication (I queried them to gain from their wisdom);
b) the group already knew they were completely dependent on voice repeaters to accomplish their perceived service goal of providing communications to the hurricane shelters.

At first convinced the EOC was the place to begin volunteering, I spent many hours studying and objectively measuring the HF capabilities of the EOC equipment, and wrote up detailed suggestions for improvements and options for new antennas. However, government tends to move at its own pace, and I soon turned attention to the most precious assets of an emergency group -- the **volunteers**, their equipment, and their skills.

ASSESSMENT OF NEEDS
- Boost existing members into higher licensure status, giving them a larger knowledge base in the process.
- Provide a backup voice communications option if the existing repeaters quit working, for whatever reason.
- Suitable HF antenna for the EOC station.
- Suitable HF antenna for the Red Cross station
- Much higher VHF antenna for the EOC station (which

geographically was located at a relative low spot)

- Digital capabilities at both EOC and Red Cross base stations.
- Portability for both Red Cross and EOC stations.
- Backup power at the Red Cross base.
- Backup digital repeaters for the one exiting repeater
- Digital skills throughout the ARES group.

PROCESS

Then began the ticklish process of encouraging local amateurs with more actual "emergency" experience than I, to gain additional skills, assets and strategies, while I simultaneously worked to leverage their experience to educate myself. In their eyes, their handheld and mobile VHF rigs had been completely adequate for all previous emergencies in our areas (typically just 60 mph winds from glancing hurricanes to our inland location), although they recognized their dependence on voice repeaters. They had not really given much thought to a Coronal Mass Ejection (CME)[17], Ted Koppel's cyber-grid-attack motif, an extremely destructive tornado, extremely destructive hurricane, EMP or biologic attack, novel respiratory virus event or anything similar. Nothing where vast amounts of information would badly need to be transmitted over *completely broken* existing systems. Yet those are exactly the type disasters when their skills and equipment would be most crucial.

> It was as if the Army had trained
> only for small civil disturbances.

Participating in the club's Field Day exercise helped me see more of their strengths at HF--- but I was the only person doing any digital.

Step by step, the list of "needs" penned above became **visionary goals** for the local group. Every chance I got, I brought up those goals and turned them into friendly challenges that garnered increased interest from most of the group.

> Really valuable volunteers respond very well to friendly
> challenges to grow their ability to serve others.

Teaching two Technician license classes helped by bringing in additional hams who trusted my guidance and were eager to learn.

Renee got me an opportunity to speak to the in-progress local CERT Training.[18] I took enough radio equipment for two VHF stations--a complete packet WINLINK server as well as a client station, and ran both right in their auditorium, moving emails across the training room with intriguing audible radio squawks and bleeps --- and gained 2 strong radio students (Adam & Celeste) (on the spot!) for an upcoming Technician Class.

We started working on increasing licensure status with projects designed to increase knowledge base and add assets. A new HF antenna went up behind the Red Cross -- but was far too low to the ground to function for anything other than convincing everyone that a higher antenna was necessary. A Slim-Jim[19] VHF antenna-building Saturday introduced the concepts of weekend education sessions, building your own equipment, and even some Smith Chart explanations of the matching section of a Slim Jim.

Another Saturday went toward putting up a higher, non-resonant, balanced-feedline type HF antenna.[20] Most of the group had never seen that sort of antenna and were fascinated to see it work. Of course, it was immediately used to demonstrate digital WINLINK contacts!

With so many in the local area limited by their license and/or equipment to VHF frequencies, a way had to be found to give them VHF access to WINLINK services, especially in the event of a disaster. The solution was to add a Windows-based RMS_PACKET packet server to my WINLINK station and then begin to help people learn how to make packet connections, at first using the popular (but pricey) Tigertronics Signalink. [21]

> The linbpq software for Raspberry Pi, by John Wiseman G8BPQ, is one of the most valuable assets to digital emergency communications, after WINLINK itself.

When I found the wonderful Raspberry Pi-based `linbpq` node software generously provided to the ham world by John Wiseman G8BPQ[22] --- this was a huge advance. Fully working, used Terminal Node Controllers are scarce, and new ones are expensive. Learning how to build very inexpensive VHF Node stations was a key goal of mine that `linbpq` would make possible.

That allowed our group to begin to put up **residential digital node stations** that would perform complete digital repeating using the `linbpq` software on an inexpensive Raspberry Pi single board computer. Renee and Joe were the first volunteers to have a station installed at their houses.

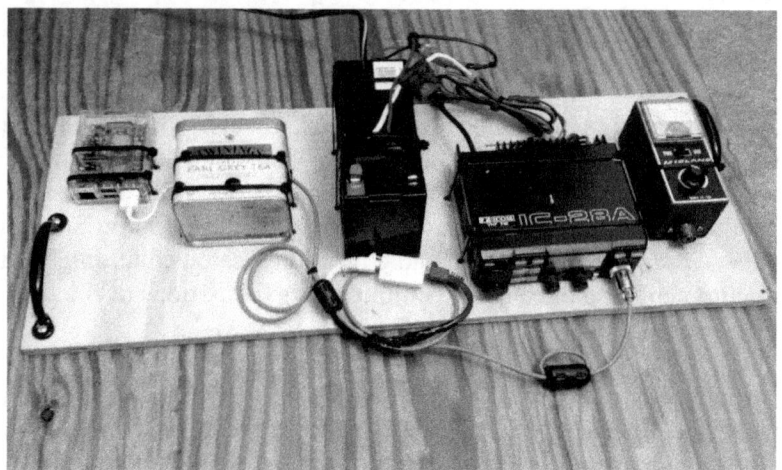

Figure 3-1. *Residential digital repeater station, that can be moved easily and also used for voice communications. From the left: Raspberry Pi; homebrew digital interface circuit (in a Tea tin); 7Ahr battery backup; used Icom-28A transceiver, used SWR meter used for relative power output and embedded gas discharge protected.*

An extremely valuable skill (that a friend Sam R. taught me) made the antenna part possible--- how to use a slingshot, 2 ounce fishing weight, and braided lightweight 80 lb. test fishing line[v] to put a line through a tree almost anywhere desired If you don't know how to do this, it was of inestimable value to me in helping people gain height for their VHF antennas. (In a disaster, I should easily be able to put ropes through non-working towers at 100+ feet height and hoist up emergency antennas)

With only a few slingshot attempts, the homemade SlimJim antennas went up 45-55 feet at both Renee's and Joe's houses and we had our first two residential digital node stations on the air! And happily,

---

v   The only difficult item to find is brightly colored 80 lb. braided fishing line. I recommend this source:
https://www.amazon.com/gp/product/B00MGA5VD8/ref=oh_aui_detailpage_o00_s00?ie=UTF8&psc=1

they could reach each other!

Figure 3-2. *In this closeup, only the rooftop & chimney of a volunteer's house are visible -- the antenna is 50 feet up, just at the limb at the top of the picture.*

That was so much fun that I introduced the idea of "antenna raising parties" (similar to old-fashioned barn-raisings) and we put up an incredibly complicated multi-band HF vertical for Renee.

From each station I learned more techniques to streamline the process.

Reducing costs at every possible step made it possible to put up a great network at minimal cost. Ebay used transceivers, inexpensive Raspberry Pi's, homebrew interfaces -- the more you learn, the cheaper it becomes.

At the same time, we continued frequent ARES training sessions at monthly meetings, thanks to the helpful attitudes of Charles, the local

ARES leader. Charles is smart as a whip and one of the most patient people I've ever met -- time after time during training sessions he would take over when I became frustrated trying to explain something to one of our volunteers. Progress was being made!

Signalinks are relatively expensive and a $5 Adafruit sound dongle[23] currently interfaces better to the Raspberry -- so we bought parts and held two weekend "solder sessions" where we actually started nine interface circuits -- and eventually finished six! Renee graciously hosted these marathon circuit-building sessions. *This was the first actual circuitry building that many of our local hams had ever done.* Since most volunteers are retired, we ran into problems with presbyopic vision, minor tremors, and other little obstacles that took patience and individual help to overcome. But it gave us a chance to go over circuit design and many practical details that are important to hams who have to be resourceful during emergencies.

Figure 3-3. *Homebrew digital interface with Adafruit sound dongle, transformer isolation, relay output, housed in a common 4" electrical utility box. Later constructions were a bit more tidy.*

Two recent Technician class graduates, husband and wife team Adam and Celeste, were next on the list and their antenna went up easily. A new problem however -- huge levels of broadband interference were immediately heard when we hoisted their Slim Jim up to 50 feet. That led to the homebrew construction of two VHF 3-element beam antennas, and all kinds of interference-hunting efforts --- and in the end it turned out to be some faulty lighting power supplies right at Adam & Celeste's house! So much was learned however, even through the solving of this temporary setback.

Meanwhile Tom Cox PhD RN[vi]--- NK3F -- enthusiastically agreed to have a station installed at his house. He had his own fishing pole-based strategy for placing lines high up in pine trees, and he was incredible adept! Tom had some previous Linux experience, and came up with ways to log every station his packet repeater heard, and took great delight in demonstrating DX contacts (via packet) that were way beyond anything we previously thought possible.

> One-on-one working with individual hams proved to be generally much more powerful than classroom demonstrations, though both probably are necessary.

With all these stations, and an improving understanding on my part, we were able to construct a series of Node stations that recognized each other, formed Routes in their memories (similar to the way MESH networks form their own paths) and functioned amazingly well to move email traffic throughout the full length of our city. "Chat" sessions, which can be used to further "common operating picture" between officials, are also easily held, albeit at 1200 baud speed. Then, just as I had been concerned, the local high-powered digital node station, W4DFU-7, began to develop problems! [At press time for the First Edition, its antenna has fallen over...since then we not only repaired it but added an entirely new radio system, developing a dual-frequency node.] Soon we were more dependent on Tom's incredibly wide-ranging NK3F-7 station and it was doing admirable service for us. Adam & Celeste's machine would be a good backup. Both stations are scheduled for upgrades as our group gets more advanced.

Along the way we developed contacts with the ARES group in the next city south of us, building a win-win cross-pollination of their skills and ours that still continues.

The Raspberry Pi Node stations were absolutely key to helping our group gain radio knowledge. They gave our members inexpensive

---

vi He's not only a very smart and friendly guy, but he was willing to have his real name used in this book.

chances to dramatically increase their knowledge and competency levels in so many areas:

- Understanding of antenna performance
- Antenna matching techniques
- Transmission lines
- Emergency antenna deployment techniques
- Basic LINUX commands
- Understanding of TELNET and PuTTY
- Basic understanding of TCP/IP and security (we had international bots attempting to break into our systems at one point)
- Tremendously greater understanding of "classic packet"
- Grasp of SWR
- Understanding of link budgets and communications links

Along the way, Renee noticed unusual network activity to her Raspberry and we got a lesson in the risks of using default, commonly known ports to access our Raspberries over the internet via port-forwarding through routers[24]-- we were being attacked by "bots" from China! A simple fix to move the ports to unusual addresses fixed the problem. In a later chapter specific guidance will be given.

BIG BREAKTHROUGHS

As we gained more and more contacts with surrounding emergency communications groups, I was challenged to take a FEMA course of how to set up drills, workshops and full scale exercises, which I did-- and it was a game changer!   Hurricanes are the big well-known risk in Florida (though widfires and plenty of other things can happen) --- and each May it is customary to hold a training simulation.   This year (2017) ours is a HUGE full scale exercise, involving 4 locations in our county, connections to one or more other groups outside our county, and hour-by-hour scenario changes delivered by secret envelope to each location. Talk about challenging!

This level of commitment on our part was almost immediately rewarded --- a simple phone call to the Florida Forestry Service (a part of State government) and we were unexpectedly given full access to an unused VHF antenna on a high fire tower (fed by 7/8" heliax!) --- and boom! our coverage of the western side of our county, and into counties west of us suddenly exploded! Further, we got a gracious offer of temporary use of a mobile crank up tower! Our Emergency Management office became very interested and far more involved in our efforts as a result. Almost 30 pages of protocol and emergency communications information were developed and placed on a simple web page along with a couple dozen more educational articles we've developed.

Our Hurricane full scale exercise is very aggressively ambitious. In a real disaster scenario with loss of normal communications, ham radio operators will be asked to handle scores to hundreds or more messages --- overwhelming tasks. Rather than just sending a handful of messages in our full scale exercise, we set out to send dozens. I think that sends a message that gains your group the resources you need to really make a difference. *I can't emphasize enough that building a quality group, setting meaningful goals to actually accomplish disaster communications, and arranging for objective and candid evaluation of our own performance, pays big dividends.*

Although we are still quite a fledgling group, we've made significant progress even beyond where we were when the first version of this book was published. At that time, I estimated we had "more than 20 miles' worth of a linear string of VHF digital repeater connections leading to a VHF/HF full service Winlink server".....Excellent coverage....within an area of more than 1200 square miles..." (Figure 3-4)

With the addition of a station in an small town southwest of us, and the fire tower node station, our non-linear network of stations now spans almost 40 miles and we probably have digital coverage (to a 20 foot high antenna[vii]) of 4,000 or more square miles, probably encompasing 300,000+ lives. [25]

That is every-day reliable coverage that does not depend on any unusual propagation. During E-scatter and tropospheric ducting, some of our group reach out 100+ miles with these simple packet stations. We now generally can link the extensive Ocala network of stations back to ourselves and to the very long SEDAN network as well. We've learned how to build dual-frequency stations (popularized by the TAPRN network) that allow us to segregate and combine networks as needed. Skill levels in our group are still choppy and variable --- but we've had a ton of fun, learned even more, and drawn significant attention to

---

vii Assumptions were residential repeater antenna height 50 feet, 25 watts, 5 db coax loss. Receiver system 20 foot antenna height, 0.5 microvolt sensitivity, 90% reception accuracy, and 12db increase in signal to have excellent coverage.

emergency communications skills in the process.

> **Our estimates and experience suggest that every extra foot of antenna height up to 60 feet is far more important than the extra loss of even inexpensive RG-8X coax. Slingshots and trees were keys to our coverage!**

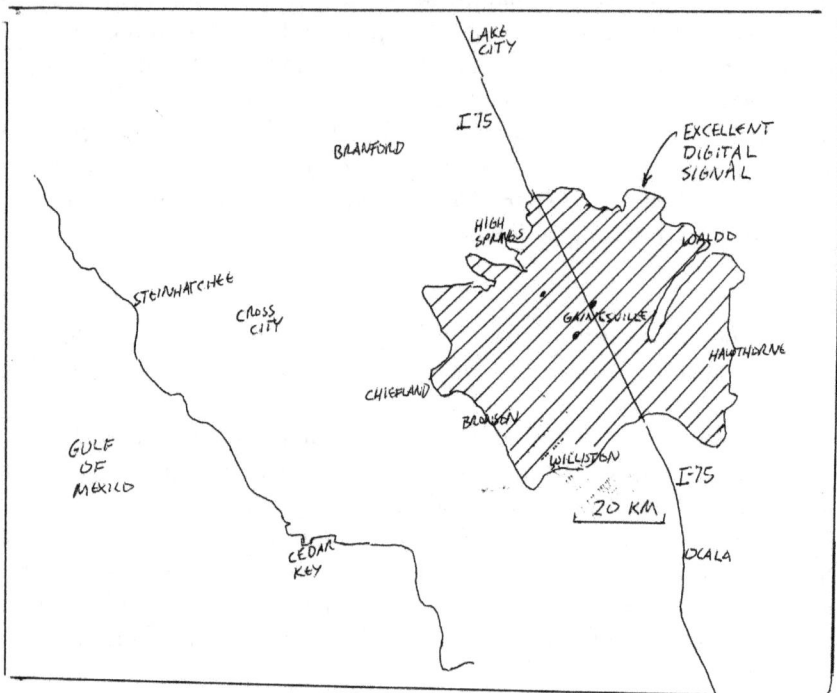

Fig 3-4 *Approximate "excellent signal level" coverage area of 3 residential digital repeater stations in the first version of this text. "Workable" range may be 25% larger in most directions.*

Fig 3-5. *Current coverage (defined as to a 15-foot antenna) now extends almost to Cross City, Lake City, and Ocala, significantly larger coverage area than just a few months ago; does not include our ability to link to Marion County network south of us.*

So the infrastructure -- the resilient network our volunteers can pick up and move at any time -- is in place. Our volunteers are starting to get the client side figured out as well. Several can now not only do "classic packet" connections and chat, but can do WINLINK email over a VHF network either direct, digipeated, or via connection scripts. We now have two additional WINLINK sysops, albeit VHF for now. As you'll see in later chapters, we now have nascent ability to do all the types of important communications listed in Chapter 2.

Figure 3-6. *Inexpensive plywood-based digital/voice client "go-box" station, including used Icom-28 transceiver, digital/microphone combiner box, Signalink, and 120VAC inverter. 7AHr gel cell is at the rear of the lower platform. USB cable to go to portable computer is coiled at the left.*

Your group can do this also. The following chapters will attempt to give you the details and save you some of the trials and tribulations that we went through!

PUTTING OUR PROGRESS IN PERSPECTIVE

Our group in Alachua County, Florida, has focused almost exclusively on *low-cost* digital & voice emergency communications systems. That meant that we chose to use decades-proven AX.25 Packet for VHF digital, using Bell 202-type modulation (two tones, 1200 Hz / 2200 Hz). This made it very easy to use almost any FM VHF transceiver. For HF WINLINK, while members of our group possess expensive SCS PACTOR modems, we started -- and still support -- inexpensive soundcard-based WINMOR.

The extremely low-cost route that we have pursued so far is not the only option, of course. With more direct connections inside VHF radios (not just through mic and speaker wiring) 9600 baud packet is possible and used by many communities and hams. It certainly is faster!

Newer technologies are also available. D-STAR has a total bit rate at 2-meters of 4800 bit/second, but 3600 of this is allocated for voice, leaving only 1200 bit/second for data. [26] My understanding of the data connection is that it is relatively primitive, and that additional software is required to achieve significant usefulness, for example D-RATS. [27] This technology will improve and mature over time and may become of greater interest. I'm hoping to find a KISS interface to D-STAR.

Yaesu's System Fusion[28] may offer faster digital connections (9600 baud) at 2 meter frequencies. We haven't yet pursued that route, primarily due to cost considerations, but if this technology becomes much more commonplace, one can expect the price to drop.

Broadband high speed data networks, often using tcp/ip based communications, have been developed under several names and in several locations, primarily where long expanses of "free space" is available to allow microwave transmissions to achieve long distances.[29] [30] [31] In the chapter "MESHING It Up," there's an introduction to this modality—and how to integrate it with the systems we've already developed.

Once you have the basic understandings, the sky is the limit to what you can accomplish.

4   WHAT REALLY IS THE MISSION?

This chapter represents my personal opinion.   I'm not an official of the ARRL, or the Red Cross or any other major group, so I think I can candidly express my opinion.   Of course, it may be right, wrong, or anywhere in between.

Over and over again when speaking with knowledgeable and experienced amateur radio emergency communications leaders, the concept of "communications requested by the served agency" comes up. This makes sense to a point -- certainly you wish to develop relationships with governmental & disaster agencies, and certainly you wish to provide them with the information and communications that they need and prefer, and not overburden them with extraneous projects or goals that don't really serve their purposes.

The problem comes when this viewpoint serves to limit a group's expertise and capabilities to such a narrow spectrum that they are basically of no service beyond just that narrow set of "communications requested by the served agency."

There is no law that says you cannot gain more knowledge and wisdom in amateur radio. No law that says you can't continue to increase your skills, assets, your protocols, your abilities and strategies...

Because when a REAL disaster comes, there is virtually no limit to the communications needs, when all normal communications are wiped out.   Police, fire, hospitals, ambulances, banks, businesses --- no one has the communications they need to preserve order, rescue imperiled citizens, deal with prisoners, food trucks, those injured, the chronically ill etc. Obtaining fuel, water, food become prime needs. Communications to arrange for requests, responses, logistics, delivery, dispensing --- these become paramount.

You'll probably be overwhelmed no matter what your developed capabilities are --- but your community will be very grateful for every bit of ability and asset and backup technique you and your group have developed!

A little vignette clearly illustrates this point. I was privileged to attend an "after-action" report at a local Emergency Operations Center after a "near miss" with a really scary hurricane. There was an amazing amount of complaining from almost all representatives at the meeting, that they could not get information and could not disseminate information, as well or as fast as they needed or wished.

Yet not a single communications system had failed during this "disaster!" The phones continued to work, the cell towers functioned fine, the Internet was still up, all the utilities still functioned. And still they reported frankly that communications was a real problem.

Since many hams in the community held to the concept of developing only those communications "requested by the served agency" (and as a result had a relatively immature emergency communications system) --- I voiced the simple question, "What will you do in a real emergency when the normal telecommunications DO FAIL?"

The gentleman in charge recognized the irony and answered that essentially, they would be looking to the amateur radio operators in that meeting and in our community to do ALL THE COMMUNICATIONS!

So, in reality, the "requested communications of the served agency" was, in their minds, limitless. Anything that quit working, they hoped we would be able to back up! The gentleman was being honest!

I bet the same is true in many communities across our nation. Ham radio operators have inappropriately limited themselves to providing communications to this or that shelter building, by voice, using repeaters, from one particular government building. When in reality, if a real disaster happened and all communications quit -- which has happened in recent history several times in the United States --- the "served agency" would suddenly wish that we could provide communications to

- state agencies
- other counties with mutual aid agreements
- hospitals
- ambulance dispatch systems
- fuel delivery sources
- prisoner transport systems
- water & sanitation systems
- utility companies, of all kinds
- state government and possibly even federal government
- an enormous number of ESG groups
- and the list just keeps growing....

One only has to read the accounts of the difficulties and responses to major disasters such as Hurricane Katrina to realize that every bit of communications ability that your group can develop might be called upon and then some, at some point in your community's future.

Thus while you might just be working on a simple communications system with one or two "served agencies," your leadership should be thinking much, much larger, and planning for assets and skills that will serve your community if a REAL disaster hit the area.

In Katrina, it is reported that in one 24-hour period, MILLIONS of attempted phone calls failed. Modern administrative management involves copious lists, spreadsheets, databases of everything imaginable. If your group's communications skills can't deal with digital "record" traffic including computer files and spreadsheets; if your group's idea of a "full scale exercise" is sending a small handful of messages, then I would suggest you rethink your goals, and your training level....while you still can.

The basic point of this chapter is: even if your "served agency" sees only a very limited role for amateur radio, you should still train for a much wider role.

So....it should be obvious that you clearly need to be an expert at the communications you DO know will be needed---and then some. One of the reviewers of this manuscript pointed out the need for training and skills – just as you would not expect to have to train an electrician hired to put an outlet in your house, the agencies shouldn't have to expect to train you in how to do radio communications!

You can't just expect to pass your ham radio license, purchase an off-the-shelf handi-talkie and think that you are the cream of the crop of emergency communications! That should be obvious. This book will serve to provide several areas where mastery of your craft will be invaluable--- digital communications, antennas, propagation, email systems, file transfer, not to mention understanding how to operate your equipment, and how to function in a radio net, or WINLINK system. Reading is only the beginning – just as you'll read more about the actual full scale exercise that our group has worked you, you'll need to get actual experience, and the more of it the better, in either simulated or real situations, to develop real expertise. .

5 THE BASICS THAT APPLY TO EVERYONE

There is a certain amount of basic information and preparation common to all amateur radio emergency preparation. While most of this book is geared to some fairly esoteric communications assets & skills --- which I think will be absolutely invaluable if another disaster anywhere close to the size of Katrina occurs --- there are certain basics that need to be attended to, even if your local disaster can be handled with only a few handi-talkies and a day or less of your life.   So let's get those out of the way.   They fall into these categories:

1. Requisite training to be accepted by government organizations
2. Basic information on local communications assets
3. Personal preparation

Requisite training to be knowledgeable
You'll be in a significantly better position if you will take the free online courses from FEMA listed in the table below, and keep multiple hard-copies of your certificates from each course. Also sign up and take the ARRL introductory EC-001 course, which has a cost. (And in some circumstances, without basic FEMA training, you would not be able to volunteer.)[viii]

Follow the link in the text on this page to obtain a FEMA Student Identification Number: https://training.fema.gov/is/crslist.aspx

---

viii Personal Communication, Steve Waterman

Course Title	URL
IS-700.A: National Incident Management System (NIMS) An Introduction	https://training.fema.gov/is/courseoverview.aspx?code=IS-700.a
IS-100.B Introduction to Incident Command System	https://training.fema.gov/is/courseoverview.aspx?code=IS-100.b
IS-800.B National Response Framework, An Introduction	https://training.fema.gov/is/courseoverview.aspx?code=IS-800.b
(ARRL) Introduction to Emergency Communication (EC-001)	See link on this page: http://www.arrl.org/online-course-catalog

Information on Local Communications Assets

Fill in as much of the following information as you can. The ARRL Net Directory may be of help: http://www.arrl.org/arrl-net-directory-search

Net Name	Frequency	Local Days	Local Time
Section/Local Net			
Section/Local Net			
Section/Local Net			
Region Net of interest			
Region Net of interest			
Area Net of interest			
ARES local net			
Other non-NTS net			

Document Local Assets

Local Repeater	Frequency/Offset	Tones	Comment

Important local contact information

Contact	Phone #	Email
1.		
2.		
3.		
4.		
5.		
6.		
7.		

Important Local Emergency Information

Contact	Phone No.	Comment
LOCAL EMERGENCY #:		
POISON CONTROL		
Police non-emerg.		
Fire non-emerg.		
Electricity Supplier		
Gas Supplier		
Local Airport		
Local Airport Tower		

Red Cross		
E.O.C.		
City Hall		

Family Emergency Arrangements

Person/Item	Contact	Plan?	Other

My Radio Equipment/Assets:

Gear	Location	Extra Power?	Other
1.			
2.			
3.			
4.			
5.			

Personal Items for Deployment

ITEM	STORED	Other
1.		
2.		
3.		
4.		
5.		
6.		

Medical Information

ALLERGIES (list all medications & what happens)	Item 1_____ 2_____ 3_____ 4_____ 5_____ 6_____	Reaction 1_____ 2_____ 3_____ 4_____ 5_____ 6_____
MEDICATIONS (list all meds, dosages)	Medication 1_____ 2_____ 3_____ 4_____ 5_____ 6_____	Dosing Schedule 1_____ 2_____ 3_____ 4_____ 5_____ 6_____
Known Medical Problems	Describe	
Personal Physician	Name	Phone / Hospital
Specialist Physician		
Previous Surgeries?	Type	Approx Year
Other Important Information		

Important Numbers

ITEM	Information	Other
Social Security #		
Driver's License		
Automobile Ins.		
Health Insurance		
Bank Account		
Credit Card		
Credit Card		

Important Documents

Location of WILL	
Location of financial information	
Next of Kin Contact	
Other Kin Contact	
Healthcare Surrogate	
Living Will Information	

Suggestions for your personal go-bag:

- Flashlight
- Pocket knife
- Cell Phone spare charger
- Cell Phone backup battery

- 2 meter/other handi-talkie
- Extra Batteries / Charger
- Comfortable Headset/Earphones
- VOM meter
- Small Tools
- Basic soldering iron, solder
- FRS radios (2)

If you are digitally equipped or provide HF WINLINK services:
- WINLINK radio equipment
- Appropriate emergency antenna
- TNC/Interface
- SWR meter
- Antenna Tuner if required

- Map for your area
- Compass
- Canteen with water
- Matches/Lighter
- Hand Sanitizer
- First Aid Kit
- Whistle
- Clipboard
- Sunglasses
- Sunscreen
- Hat
- Writing Utensils
- Notepad
- Drinking water tablets
- Any prescription items
- 550 cord

- Earplugs
- Duct tape
- Lightweight rain gear

- Appropriate clothing/coat for the time of year
- Food as appropriate for deployment envisioned
- Water as appropriate for deployment envisioned

6   PORTABLE CLIENT STATIONS

This chapter will go over how to build relatively low-cost voice and digital client stations for FM-VHF & also those for HF access. These are called *client* stations because they allow your volunteer hams to use voice and digital transmissions to send and receive information, and aren't primarily designed to relay information for others.

They need to be PORTABLE because your volunteers may need to serve at any of a large number of locations, including shelters, government buildings, hospitals -- and these may not even be in your home city.   Having all the gear nicely assembled on a rigid form makes moving quickly a lot easier.   To reduce costs, we used simple plywood (roughly 1/2" thick), either as a single flat platform, or with 1"x4" or 1"x6" sides, as a box effectively giving two "shelves."

To use voice you really only need a microphone and speaker/headset. To take advantage of digital broadcast or error-corrected digital email, however, you'll also need a computer, appropriate software, and either a TNC or interface circuitry to go from your mic/speaker signals into the computer.

Because of the differences between the background noise, interference, band propagation and emissions between VHF and HF access, this chapter is broken into four sections:

- VHF Client Station Construction
- VHF Client Station Usage
- HF Client Station Construction
- HF Client Station Usage.

49

Microwave MESH systems will be covered in a later chapter.

VHF CLIENT STATION CONSTRUCTION

There's more than one way to accomplish the digital connection, as shown in Table 6-1.

TABLE 6-1   Digital Connection Techniques (VHF)

Method	Hardware Required	Software Required
Classic Packet using a hardware TNC	TNC such as Kantronics KPC-3 or MFJ TNC-X	Simple terminal emulator software
Classic Packet using soundcard technology	Due to fast back&forth error correction, you'll need some sort of automated push-to-talk control. Signalink or homebrew vox with fast response. (Note: this is a correction from V1.0 of this text.)	UZ7HO has a great solution with easyterm38 (zip file distribution) connecting by AGWPE interface to soundmodem97 (zip file distribution). Works very well and simple to use.[32] **STRONGLY RECOMMENDED** MixW[33] is a full featured (commercial, $) windows-based software that can do packet via soundcard. Free trial versions are also available.
Soundcard-based broadcast (less common for QSO's, but very useful for bulletins)	Tigertronics Signalink, or other interface to either external or internal computer soundcard	FLDIGI, Ham Radio Deluxe, MixW or similar will work fine for this.
WINLINK	TNC or Signalink,	WINLINK EXPRESS free

email access via Packet	or other soundcard type interface; appropriate computer port (serial or USB)	software; if using soundcard, add UZ7HO soundmodem.exe[34]

A simple VHF client can be built with a handheld FM transceiver capable of reaching a nearby VHF Packet RMS server, a sound-card interface such as a Signalink and a laptop computer running UZ7HO soundmodem.exe interface software and WINLINK EXPRESS client software as well as UZ7HO easyterm39 (term.exe).

To make this station portable and emergency-hardened, the radio can be mounted on plywood box or mounted in a rackmount protected rugged box such as used by traveling musicians to protect their amplifiers. Additional equipment such as a gel-cell storage battery, possibly a 120V inverter, necessary fusing or circuit breakers, possibly a charger or 120VAC power supply, and some means of output power observation and/or SWR measurement may also be included. (I recommend a fuse connected directly to any storage battery.) A simple low-cost example is shown in Figure 6-1.

Figure 6-1. *Homebrew client portable emergency station capable of voice or digital FM operation.*

MICROPHONE COMBINER CIRCUITRY: It is very advantageous for the user to be able to seamlessly switch back and forth between voice communications and sending or receiving digital information.   While the receiver audio can be sent both to the computer and also to a speaker (equipped with a separate volume control), the microphone signals from the physical microphone and the computer soundcard output need to be mixed with appropriate separation. Furthermore many current microphones are electret-based and operate at a positive DC bias voltage of perhaps 7-8 volts, whereas the output of a Tigertronics Signalink is the secondary winding of an audio transformer which would quickly short out the DC bias of the microphone and possibly damage the transceiver. A simple circuit (see below) can be used to adequately separate these mic signals.   The Push-toTalk lines from both microphone and Signalink can usually be tied together.

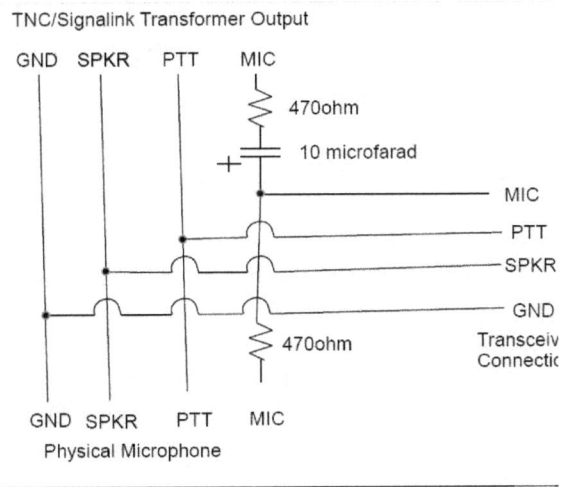

Fig. 6-2 *Schematic for microphone combiner circuitry. Combines signals properly and prevents transformer output from Signalink or other TNC from shorting out DC bias required by many microphones. Do test to find out if your mic stays "live" during data transmissions, and if so, consider muffling or disconnecting it if using on HF bands where voice would not be appropriate on portions of bands.*

Parts:
> 470 ohm resistor, 1/4 watt
> http://www.digikey.com/product-detail/en/stackpole-electronics-inc/CF14JT470R/CF14JT470RCT-ND/1830342
> 10 microfarad capacitor:
> http://www.digikey.com/product-detail/en/nichicon/UVR1E100MDD6TP/493-12772-1-ND/4328365

Note that some radios allow digital to be connected to a special accessory jack, and automatically handle digital separately from analog mic – these radios wouldn't need this circuitry.

SPECIALIZED CABLES

Every radio has its own microphone connector and there is no standardization. In order to deal with this, Tigertronics provided two levels of customization: a plethora of different microphone cables designed to accommodate different transceivers, and additionally a dual-inline (DIP) jumper socket inside the Tigertronics Signalink that would allow the four relevant signal lines (ground, microphone, push-to-talk and receiver audio) to be positioned as needed on the pins of a RJ45 female modular socket on the back of the Signalink. The MFJ 1204 has a similar system.

In an active ARES group, it is likely that flexibility may be needed as different equipment is put into service and moved around. Having such inherent variability in the Tigertronics Signalink pinout is a recipe for problems. As a result, our group in Alachua County has standardized on the following RJ-45 wiring on all interfaces, regardless of radio:

Pin **Signal**
1   Microphone audio to be transmitted
2   Ground (both push to talk and microphone)
3   Push to talk (gets grounded in order to cause transmission)
4   Not Used
5   Receiver Audio (signals to be decoded)

NOTE: **Various manufacturers number the pins of the RJ45 in different order.** Except where indicated otherwise, in this book the numbering will be as shown in Figure 6-3, which shows a RJ-45 plug with the contacts facing the viewer and the cable end toward the viewer.

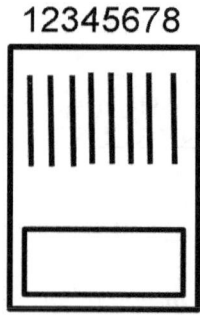

Fig. 6-3 *RJ-45 plug pin numbering.*

There are unfortunately TWO color-coded standards for wiring RJ-45 connectors for use with networks, T-568A and T-568B.[35] MOST CABLES are wired in the T-568B standard, and this is the standard utilized in this book. The color codes for the pins in that standard (and the signals for which our group uses them) are shown in Table 6-2:

TABLE 6-2 T-568B COLOR CODES AND OUR SIGNALS

Pin	Color	Alachua County ARES Signal Usage
1	white-orange	mic
2	orange	ground
3	white-green	push-to-talk
4	blue	unused
5	white-blue	speaker audio
6	green	unused
7	white-brown	unused
8	brown	unused

Every transceiver used in our group has a custom-made audio cable created so that it can connect to a standardized Signalink RJ45 socket presenting signals as above. Where possible, that audio cable uses shielded cable, to reduce radio frequency interference possibilities.

HOMEBREW SOUNDCARD INTERFACES

Furthermore, our group has made homebrew "$10TNC" Signalink-equivalents, as described in a later chapter. In order to continue the flexibility we wanted those to present the same pinout as our Signalinks. At that time, we didn't find an easy-to-use female RJ45-socket that was hardy (they do exist). Thus we have typically wired our homebrew interfaces with cable terminated in a RJ45 plug, (pinout as above) and then utilized a female-to-female inter-connector so that the $10TNC effectively presents a female RJ45 socket with exactly the same pinout as all our Signalinks are wired.

VHF CLIENT STATION USAGE

Just as Table 6-1 suggested, there are multiple modes in which a VHF client station can be utilized, to accomplish different goals. Here I will discuss "classic packet" first, then "broadcast modes" and finally WINLINK email operations.

A. "CLASSIC PACKET" VHF OPERATIONS

UZ7HO's free software easyterm39 (term.exe) makes digital packet QSO's within reach of any ham with a computer, a VHF transceiver, and the ability to put together a push-to-talk electronic control (see later in this text). AX.25 allows (and has for a couple of decades or more) error-free digital QSO's that can traverse a large number of node and/or digipeater stations. Test messaging on cell phones is popular for exactly the same reasons that packet was originally so popular. Packet contacts allow "record" error-free emergency communications to pass

- tactical messages
- bulletins
- files, even including ICS messages

and it occurs almost automatically! A YAPP-based transfer of a file requires no manipulation on the receiving end -- it is simply started at the transmitting end! These kinds of abilities are perfect for shelter and other emergency communications where a system can simply left "on frequency" and the local EOC or any other party with urgent information can easily reach desired recipients (albeit one at a time) and put important information on their screen, or transferred by files. It only requires the recipients system to be on frequency, and reacheable.

As a demonstration of "classic packet", here's a brief transcript (copied from my computer screen) demonstrating several features of bare-bones keyboard-based packet radio conversation (with my transmissions **bolded**):

```
--------------------------------------------------------------------------------
    ?
 JNVILL:KX4Z-7} CHAT RMS RELAY CONNECT BYE INFO
NODES PORTS ROUTES USERS MHEARD
 mh 4
```

```
JNVILL:KX4Z-7} Heard List for Port 4
W4DFU-7    00:00:02:45
NK3F-4     00:00:04:35
NK3F-7     00:00:04:57
C 4 NK3F-4
JNVILL:KX4Z-7} Connected to NK3F-4
[BPQChatServer-6.0.13.2]
NK3F's Chat Server.
Type /h for command summary.
Bringing up links to other nodes.
This may take a minute or two.
The /p command shows what nodes are linked.
KX4Z   : gordon *** Joined Chat, Topic General
1 Station(s) connected:
KX4Z   at LINBPQ   gordon, ?_qth [General]
Idle for 0 seconds
Hi, this is an example of a chat session over
vhf "classic packet".     If other......
```

Fig. 6-4: *Transcript of short keyboard-based packet radio communication. What I typed (and transmitted) is **bolded** to make it more clear.*

Here's a blow-by-blow explanation so this makes more sense:

1.   Typing a "?" mark into my computer terminal tells the station to which I'm connected, that I want a list of their options/commands The packet node responds with its call sign and a standard list of options:

```
CHAT RMS RELAY CONNECT BYE INFO NODES PORTS
ROUTES USERS MHEARD
```

(usually only the first letter or two is really needed to exercise these commands)

2.   I chose to have it list the stations it has heard recently on its 4th port (which on this station I know is connected to the 2 meter rig --- for many nodes, you won't need to give the port number; it will be automatic): "**mh 4**"

3.   The node responds with a list of stations it has heard recently, and

how long since they were heard:

```
W4DFU-7      00:00:02:45
NK3F-4       00:00:04:35
NK3F-7       00:00:04:57
```

4.   I respond by asking for a connect to NK3F-4 (which is a direct connection to a chat server): " C 4 NK3F-4"

5. I then get the connection and can begin to type into the equivalent of a round-table chat room.

This example demonstrated communicating essentially with robots -- computerized repeater stations.   Once skill is gained in using these, it becomes simple to use them to quickly and easily reach live ham radio operators through them, using Easyterm, and have actual contacts, delivery of important information, or file transfer. Digital signals are easy to "route" giving packet a tremendous power to reach out.

B. WINLINK over VHF Packet

When you can't always be available and reachable, store-and-forward messages ("email") by radio, using WINLINK provides the solution. WINLINK allows you to send an email to a repository for a local ham, or reach way out and contact a station far outside the disaster area and get your email transferred through intact Internet connections. WINLINK works with free software and has amazing versatility. The more I use it, the more impressed I am with this system, developed over more than a decade.

There is very significant usage of VHF frequencies to reach local WINLINK email gateways using AX.25 packet. This is also a great way to reach a local HF WINLINK gateway for emergency communications, that also has a VHF access. The use of WINLINK over packet is fairly simple -- connect up either a real TNC or else a Signalink or other soundcard-based technology, fire up an interface software like UZ7HO soundmodem.exe, arrange for TCP/IP port connection to WINLINK, and then just follow the normal procedure for WINLINK connections, choosing "Packet WINLINK". Further discussion will be provided in the HF section below.

Note: The Winlink connection protocol isn't actually that

complicated. In a pinch, there is even a way to send/receive email by connecting to a WINLINK server using classic-packet keyboarding and executing commands from a command line: http://www.w5cwt.com/2016/04/winlink-on-raspberry-pi.html. Except for dealing with the password, it's fairly simple. [36] I've used that many times and it is actually quite easy.

C. Soundcard-based Broadcast Modes on VHF

These work identically on FM VHF as they do on HF, and thus will be more-fully discussed in the HF session below, because they are primarily used for QSO's on HF. On VHF, however, they have a great potential to be used in emergencies to allow fast and accurate broadcasts of vital (but relatively non-confidential) information such as weather bulletins, extensive lists of trouble areas, power outages, downed trees, medication supply needs, hospital evaluation information -- the kind of stuff that is tough to accurately get across on voice.

Because FM has such great clarity, particularly via a repeater, you can use full-bandwidth modes such as MT63-2000 (2000= bandwidth in Hz), but of course only one person on FM can use the frequency at a time! While useful for sending bulletins, etc., from a net control station or data-gatherer during an emergency, these modes aren't very common for normal contacts on VHF frequencies, and certainly not on repeaters! Be cautious of local etiquette!

HF CLIENT STATION CONSTRUCTION

Almost any single-sideband HF transceiver/station can provide valuable service during a significant emergency because of their ability to reach out across states or even nations to provide communications—without any remaining local infrastructure other than your hastily deployed wire antenna! For modes like WINLINK email where you need to have precise frequency control (within 100 Hz) and the ability to try multiple possible servers quickly, a digitally-controllable solid state rig is advantageous, but those have been around for almost 20 years now, so many are available on the used market inexpensively. I have significant experience with the ICOM-725, 728 and 718, all of which have worked well for me.

But I've made many digital contacts and even many WINLINK email transfers literally using the Heathkit vacuum tube SB-102 that I soldered together in high school! With a little attention to frequency measurement and stability, even a vacuum tube rig works just fine connected to a Signalink or $10TNC and feeding a modern laptop computer!

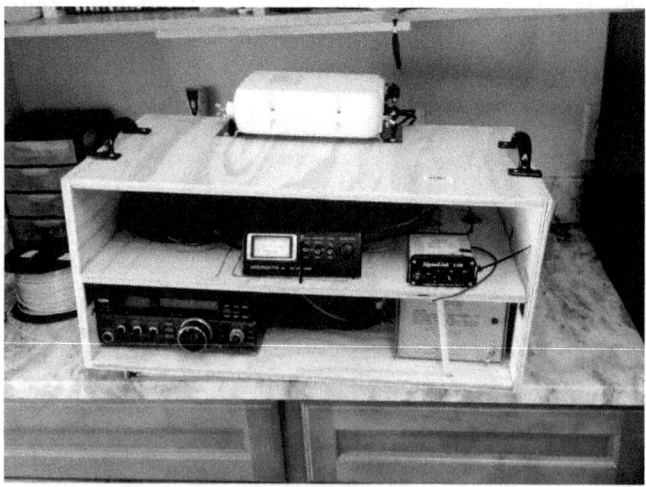

Fig. 6-5. *HF Client Station in home-made go-box. The white container on the top is an ICOM automatic antenna tuner. A VHF rig can also fit in, using the empty spot at the left of the middle shelf.*

For the strongest applicability to emergency communications, an HF station should ideally be fastened inside some sort of go-box, whether of wood or a commercial rack-mount cabinet. Add in any necessary power supplies, perhaps an inverter, definitely a soundcard or PACTOR modem interface, some means of measuring SWR, and likely an antenna tuner.

Wrapping up or rolling up an HF antenna that can be rapidly deployed, and including such tools as rope, a slingshot and braided fishing line are wise moves that will make it a lot easier to move and re-establish your station should the need arise.

For most digital applications, a station with 50-100 watts output power is well more than needed. There often won't be enough left over power to spend on a linear amplifier.

HF CLIENT STATION USAGE

For many new amateur radio operators, VHF operation using a readily-available and inexpensive handi-talkie seems easier to start off with than HF operation. Our group thus has fewer HF Client stations. A nearby city has at least two HF client stations strategically positioned at hospital or similar positions. In a disaster involving communications loss, those will turn out to be very important.

HF peer-to-peer digital communications is actually very easy using free software such as FLDIGI. Many amateur radio emergency groups have built their digital transactions on "Narrow Band Emergency Messaging System" (NBEMS) based on the extremely popular--and free-- Fast Light series of software products.[37] Bloomberg, an ARRL Section Emergency Coordinator, makes the point that voice is great, until you're asked to communicate such detailed data as a roster of evacuees, required prescription medicines, or detailed directions to a disaster scene --- then digital begins to shine. [ix] The base layer is FLDIGI, which allows peer-to-peer keyboard-to-keyboard QSO's over a truly dizzying array of possible protocols. A beginner's guide is available[38] as well as innumerable online tutorials.

FLARQ - Adds error correction

---

ix Avoid sending sensitive / confidential information via broadcast digital modes that can easily be received by any station.

FLWRAP -- File encapsulation / compression
FLMSG -- Forms manager
FLRIG -- Rig control (works with FLDIGI)[39]

Emergency communications groups should universally have some familiarity with the FLDIGI set of software or comparable applications.

DIGITAL CONTACTS ON HF USING FLDIGI

You can generally find PSK31 conversations during the day at 14070 kHz (upper side band), and in the evening on 7070 kHz.[x] Somewhat surprising to previous voice-only operators, **there can be ten or more PSK31 conversations within the 2 kHz typically occupied by**

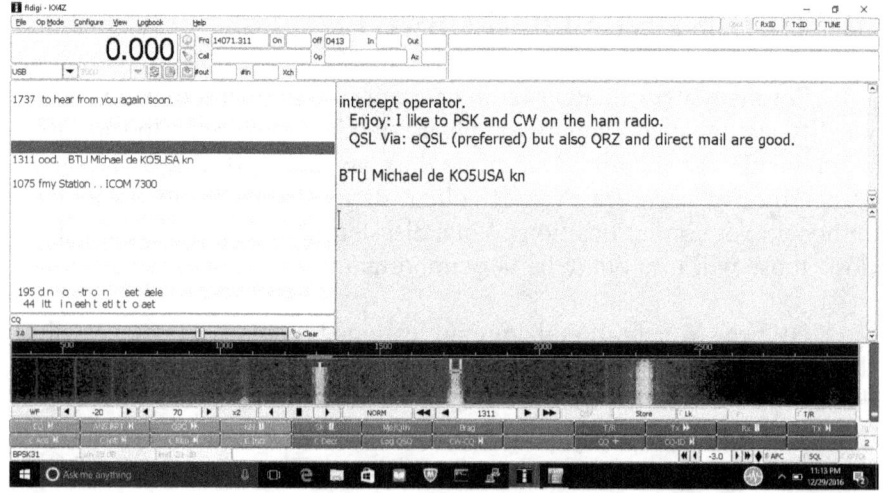

Fig 6-6. HF FLDIGI on 40 meters using a vacuum tube SSB rig. 3 signals visible in the passband of a lightly occupied frequency. Multiple conversations tracked in browser window to the left. Good copy on the selected signal showing up in the upper right hand page.

one SSB signal. Each PSK31 signal uses only a tiny sliver of bandwidth! So you may set your rig to 14070 (upper side band) and

---

x   Amazingly, if you inform FLDIGI whether you're using upper or lower sideband (selection box upper left portion of application) --- it will make things work for you no matter which you're using!

have your choice of multiple contacts, by sliding the pointer in the FLDIGI waterfall to different audio frequencies within the SSB passband. Not everything on the bands is PSK31, of course. You'll have to learn how to recognize the different sounds of different modes, because there will be RTTY, Olivia, JT65 -- all kinds of different digital modes, usually a few kHz apart.[xi]

There are innumerable YouTube videos demonstrating the setup and usage of FLDIGI on HF and VHF. FLDIGI is available for Windows, Mac and even Raspberry! You can even get it to work in a pinch by just putting radio speaker near computer microphone and vice-versa! With a Signalink between the computer and the transceiver, you'll get better copy, of course. Other programs that are popular include Ham Radio Deluxe, and MixW. Here's a nice YouTube giving practical information on PSK31 using Ham Radio Deluxe: https://www.youtube.com/watch?v=UccT47K-QoA.

After you've installed the software, configure FLDIGI with your callsign etc (CONFIGURE | OPERATOR) and set the correct sound card (CONFIGURE | AUDIO | DEVICES...check "Port Audio" and then select the Signalink for Capture and Playback).

Pick which mode you're going to use (Op Mode | PSK31| BPSK31 or any other choice). When you're ready to get your transmitter going, click into a clear spot on the waterfall, and you can use the TUNE button in the upper right hand side of the screen to send a steady signal so you can adjust the transmit audio level (on the Signalink) so that you aren't overdriving your transceiver. Don't push your transceiver too hard; 30 watts output is fine, and digital has a high duty cycle when transmitting.

Watching the "waterfall" FFT display at the bottom and adjusting the RX Volume on the Signalink, you'll soon be seeing conversations. On both PSK and RTTY, in the far left pane you can actually follow multiple conversations at once! Click on a conversation and you're on their frequency. You can also click in the waterfall to set your desired frequency.

Frequency Control
One confusion -- while FLDIGI *can* be configured to automatically

---

xi Examples of the sounds of different modes: http://www.kb9ukd.com/digital/

control your transceiver's frequency if you have a computer controllable transceiver, you can always manually tune any transceiver. It works just fine with my Heathkit vacuum tube SB-102 which certainly can't be computer controlled! In this case, you just dial the frequency you want with your rig, and ignore the big frequency display in the upper left corner of FLDIGI. I prefer to have the waterfall marked in AUDIO frequencies (CONFIGURE | Waterfall | Display; check "always show audio frequencies").

FLDIGI includes an incredible number of options you can customize, as well as up to 4 rows of soft-keys (macros) that you can pre-set . The more you use it, the more familiar you'll be. Another introductory YouTube related to digital ham radio: https://www.youtube.com/watch?v=HQIG_vMGe1A

EMAIL OVER RADIO ON HF

By far the most well-known amateur radio client email system is WINLINK[40], which can not only provide distant client-server email/attachment access in the absence of the Internet, but can also provide peer-to-peer email/attachment error-free transfer without any need for a server. And with a fair amount of confidentiality.

For WINLINK access, most client stations will initially choose soundcard-based WINMOR ("Windows Messaging Over Radio") access as the PACTOR modems are quite expensive. The same Signalink that works for VHF can be utilized for HF! In keeping with our flexibility goal, our group wires HF station audio cabling with the same pinout as our VHF stations so that everything transfers over directly from VHF to HF and vice versa.

Here's an introductory YouTube on WINLINK client operation: https://www.youtube.com/watch?v=WRv9DbqnnDg; K4REF has an entire series on WINLINK on YouTube: https://www.winlink.org/content/k4ref_how_to_video_series_winlink_winlink_express

In order to connect to a WINLINK HF RMS server, select your desired connection mode (WINMOR or PACTOR), click "connect." A new dialog box will open. If you haven't configured it before, go through the setup option to select your connection hardware. If your transceiver is computer-controllable and you've wired the appropriate connections to a USB jack on your computer, you'll want to set up the computer control options as well. The Transmit Level Test will allow you to adjust TX levels so that you aren't overdriving your transceiver -- certainly less gain than what tops it out, and usually with little to no ALC visible works best.

The Channel Selection menu option brings up a sorted list of which HF WINLINK servers are most favored by propagation at your time and place. Green color is good, yellow or red not so good. Click on any favored server. If you don't have automated rig control, dial your station accurately to the DIAL FREQUENCY (not the Center Frequency) shown to you & be sure your station is set to send upper sideband signals (even if you are on 80/40 meters). Click START and the system will begin trying to connect to an HF RMS server.

If the connection is successful, you'll see indications of the radio connection, a further Internet connection to a Winlink CMS (central message server) and then whether or not you have any messages to download. Messages in your outbox will automatically send. The system will disconnect and advice you of the speeds achieved.

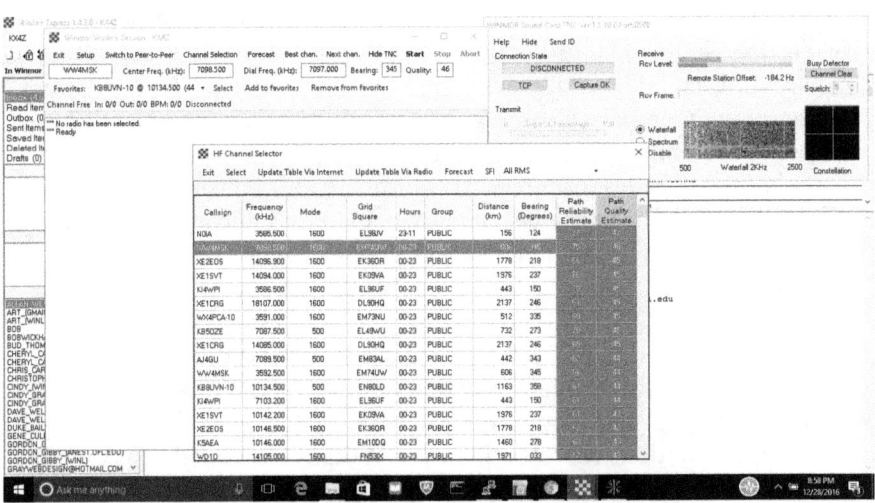

Fig 6-7 *Multiple overlapping WINLINK dialogs simultaneously. In the background is the main box for the email program. In front of that is a dialog for a WINMOR connection; in the upper right is the TNC software that implements the WINMOR mode on the soundcard device;*

and in front is a sorted list of best-possibility connection servers.

It is a finely tuned system, all written by volunteers, maintained by volunteers all over the world, and available for free use 24 hours a day, 365 days a year.

SHARED RESOURCES HIGH FREQUENCY RADIO PROGRAM (SHARES)

The U.S. Department of Homeland Security (DHS), National Coordinating Center for Communications (NCC) program for Shared High Frequency radio resources is an intentionally under-advertised program that works to provide HF radio communications between governmental units that would be impervious to cyber or other vulnerability risks. Their web page states,

> "The SHAred RESources (SHARES) High Frequency (HF) Radio program, administered by the Department of Homeland Security's (DHS) <u>National Coordinating Center for Communications </u>(NCC), provides an additional means for users with a national security and emergency preparedness mission to communicate when landline and cellular communications are unavailable. SHARES members use existing HF radio resources to coordinate and transmit messages needed to perform critical functions, including those areas related to leadership, safety, maintenance of law and order, finance, and public health "[41]

One doesn't easily find any public details much beyond that. Participation is easily available however to EOC's and other units, merely by filing an application for a license. A fact sheet explains that over 1400 HF stations representing over 100 governmental organizations are involved.[42] Periodic training is ongoing in this group.

The primary advantage to EOCs considering involvement is this offers a virtually free way for non-ham governmental officers to be able to make long-distance contact out-of-county in the event of a massive telecommunications loss, over relatively interference-free frequencies that are not in the ham band segments. The State EOC in Florida is a part of this group. The SHARES program includes digital email servers and clients, and is able to use the full speed of PACTOR IV.

There is a simple application form if your local EOC would like to have this asset in their back pocket.[43]

CLIENT STATION PROBLEMS & VULNERABILITIES

RADIO FREQUENCY INTERFERENCE

Most amateur radio operators (particularly those with HF experience) have dealt with an RF interference problem at one time or another. With digital signals, almost *every* participant will get an education in how to reduce radio frequency interference to solid state ("digital") circuity.

"RF in the shack" in years past meant you might get "bit" by the mic if it touched your face – but in the digital age, it can mean USB ports lock up, computers freeze, and digital signals from your trusty ham radio come to a screeching halt. After the issue of simply getting your Signalink or other interface device properly wired up to your radio, conquering RFI is probably the most thorny problem that new digital hams face. When you hit the TX button and everything freezes....that's a clue!

Whenever RF energy is *flowing through*, or *radiating from* the ground connections, cases, shields, and braids of your ham shack wiring or coax cables (outside surface of braid) , you are possibly going to have enough of a voltage difference between "ground" potential of one device (Signalink) and the next device (USB port) that semiconductor junctions get inappropriately activated....leading to crashes.

It is devilishly difficult to understand, because at HF or VHF frequencies, every wire, every loop, every surface can have an inductive or capacitive reactance – ground wires aren't merely wires any more, they are transmission lines!

Here are some suggestions based on what has worked for me, in operating four digital stations:

1.   **Balanced RF systems helpful**.   Fully balanced antenna systems, while not required, will be easier to deal with. Coax (at RF frequencies) has **not just two, but actually THREE conductors:** inner conductor, inner surface (skin effect) of the braid, and OUTER SURFACE of the braid (skin effect again!).   The current on the first two will likely be equal and opposite (no net radiation) – but that third conductor, the outer surface of the coax cable, can support an unbalanced current that turns your transmission line into yet another antenna, which goes all the way back to, and even includes the case of your transceiver.

A "balun" device may add sufficient impedance into this third conductor system that it damps down that current/radiation so that your RFI problems disappear.   Try a full dipole (two quarter wave wires with coax connecting the middle).   Then put a 1:1 current balun in the coax line either at the center of the antenna or as close to the antenna as reasonable.   I've used inherently-unbalanced off-center-fed "Windom" antennas successfully with digital systems – but definitely using a balun!

A 1:1 current balun is little more than 12 or so bifiliar windings of wire (or coax) with a toroidial core.   An equivalent effect is made by placing a couple dozen ferrite beads on the coax (one "turn" on each of several dozen toroids).   You can read more about building your own baluns at a howto our group has on the net: http://www.qsl.net/nf4rc/BalunHowTo.pdf

Without the balun you are almost certainly going to have some radiation from your transmission line with inherently unbalanced antennas. Use the balun to stop that from making it all the way back to your station.   (I had never previously used a balun until I started digital.) At HF frequencies, toroidal products abound for purchase. For VHF/UHF, several turns of coax in a nice pattern on a PVC form add the required impedance (more effective than "scramble wound" without a form). http://www.hamuniverse.com/w7lpnvertdipole.html

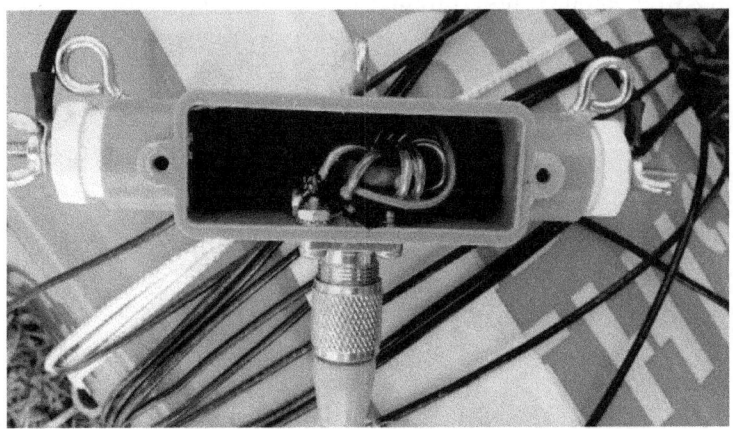

Fig 6-8. *Commercially available and inexpensive 4:1 toroidal balun.*[44] *Behind is an antenna wrapped around a 5-gallon "Homer pail."*

2.   **Shielded USB cable. The single most important thing I learned the hard way** was to use a SHIELDED USB computer cable between a Signalink and a laptop computer. Very sneaky crashes were eliminated by replacing a cheaper cord with a $5 shielded USB cable. The one I used was:
https://www.amazon.com/gp/product/B001MSU1HG/

3.   **RF ISOLATOR**   (A 2nd balun on the transceiver side of an antenna tuner.)   For HF, an inexpensive RF ISOLATOR on the coax transmission line seems to help.   These may be the bifilar wound toroid, or a bead-based product.   They aren't expensive. I've used this one but there are many others on the market:
http://www.mfjenterprises.com/Product.php?productid=MFJ-2912
You can also easily make one of these yourself-- just put about 12 turns of bifiliar (close together) wire of an appropriate gauge wrapped around a toroid such as an FT-140-61. [45] [46]

4.   **Snap-on ferrite cores** should be used liberally on all digital lines and anywhere else you can think of.   If you buy ones with larger thru-holes, you can even make a couple or 3 turns of the cable like a toroid inductor.   Here's an inexpensive set to consider:

https://www.amazon.com/gp/product/B01E5E5IY4/

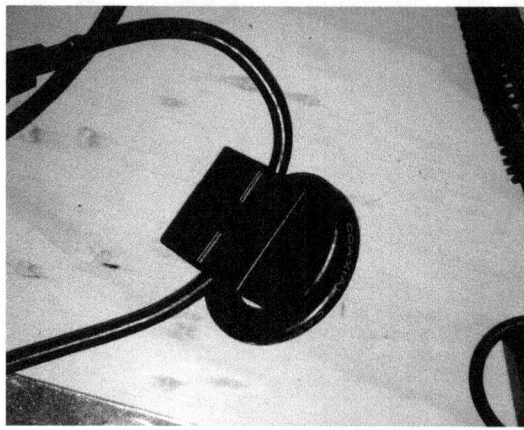

Fig 6-9. *Cable looped through a snap-on ferrite.*

5.   **LOOPS** Even simple loops ("scramble-wound") of signal or coax cables, held in shape with zip ties, can help.

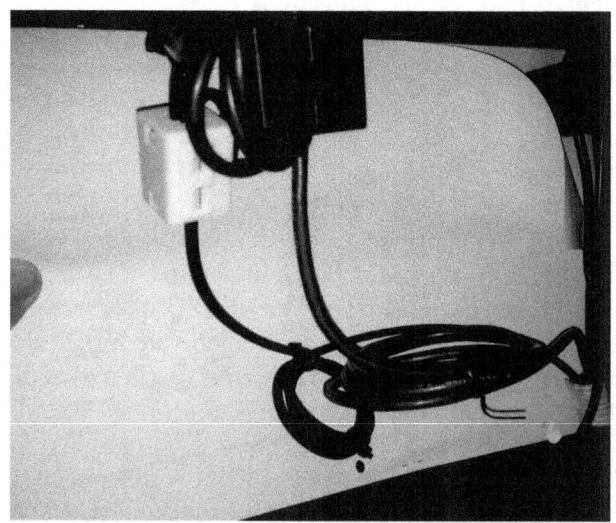

Fig. 6-10. *loops in addition to ferrites*

6.   **CAPACITORS** across DC power connections. I had a PACTOR modem that would crash....until I put a small bypass capacitor (0.01 microfarads, short leads) right at the power input to the modem. Now I put filter capacitors on any 12 volt line going to any fancy modem.

ELECTROMAGNETIC VULNERABILITY

Since the 1960's both western and eastern governments have understood the <u>electromagnetic</u> destructiveness of nuclear weapons. Testing in the 1960's indicates that solid state transceivers connected to an antenna of any length more than a few inches could be irreparably damaged in nanoseconds by a high altitude nuclear explosion (HEMP, High Altitude Electromagnetic Pulse).

That damage is caused by waves known as E1 and E2, which are electromagnetic fields approaching 50 kilovolts per meter...an incredibly strong E field. A minute or so later, alterations to the Earth's magnetic field caused by explosion products can cause extreme damage to the very long wires of a national electrical grid, potentially knocking out electrical power over vast reaches of territory for months to years. Congressionally-mandated committees have released two reports detailing the immense destructiveness of these weapons and the dramatic vulnerability of the civilian population of the United States[47] [48]

What is surprising is this vulnerability was well described by the ARRL QST magazine in the 1980's. [49] Fully functional hardening techniques which have a good chance of protecting even solid state equipment were well described. These techniques involve careful application of gas discharge tubes (which conduct above a certain voltage level) across antenna feedline terminals, filtering where possible, metal oxide varistor (MOV) surge arrestors and power line disconnects to protect against the delayed power line effects.

These are the same sorts of protections that many amateurs use to protect themselves from lightning strikes; the EMP effects have significant similarities (except for the incredibly short time course of the powerful E1 wave, far faster in onset and decay than lightning).

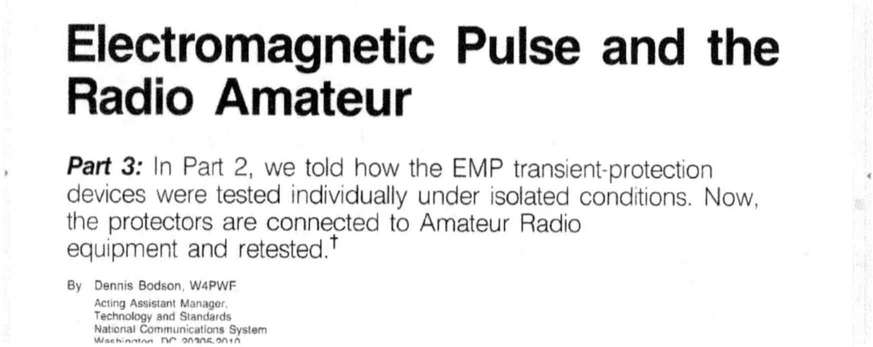

Electromagnetic Pulse and the Radio Amateur

Part 3: In Part 2, we told how the EMP transient-protection devices were tested individually under isolated conditions. Now, the protectors are connected to Amateur Radio equipment and retested.[†]

By Dennis Bodson, W4PWF
Acting Assistant Manager,
Technology and Standards
National Communications System
Washington, DC 20305-2010

Fig.6-11 *One of four articles QST published in the 1980's on EMP and simple ways to protect against it.*

It therefore only makes good sense that "emergency communications" volunteers to take prudent steps to protect backup communications equipment from EMP-induced damage. Such steps include:

- Keeping unused backup equipment inside a conducting container that functions as a Faraday shield. A holiday popcorn tin or a metal garbage can with a well-fitting lid is a good choice. Grounding is not necessary at all, but one might want to line the container with cardboard as the EMP waves may cause large voltages to appear on the inner surface. [50]

- Utilize high quality lightning protection on all external antennas that is based on gas discharge arrestors which are occasionally changed out. Alternatively, soldering in readily available gas discharge arrestors with short leads (careful RF wiring dress) where possible, such as inside SWR meters, or inside antenna tuners, on the 50 ohm side. [51]

- Reducing the bandwidth of energy from an external antenna reaching a solid state transceiver to reduce possible EMP damage by using matching systems or antennas that have inherent matching systems (J-pole or Slim Jim).

- Provide MOV surge arrestors on AC power lines.

- Provide AC disconnection in case of dangerous AC voltages possibly caused by EMP damage to split tap AC distribution transformer neutral wiring   Most higher quality UPS (uninterruptible power supplies) include this automatically.

Gordon Gibby KX4Z

7 HOMEBREW EMERGENCY ANTENNAS

Even if every Red Cross shelter, EOC building, and hospital in your county had professionally installed, commercial HF and VHF/UHF antennas, you would still need to be able to whip up an *ad-hoc* antenna at some time in an emergency situation, possibly due to damage to existing equipment, inadequate signal level, or the need to move operations to an unexpected location.

Antennas naturally fall into two groups: high frequency (3-30 MHz) versus VHF/UHF antennas, which are generally much smaller. Obviously, you need an antenna whether you are going to utilize either voice or digital modes!

HF ANTENNAS

The simplest antenna to use for HF purposes is a resonant dipole. The length, in feet, of a half-wave resonant dipole is approximately 468/f, where f is the center frequency expressed in Megahertz.   Table 6-1 gives approximate lengths for antennas on several important bands. The antenna is bisected in the middle, the two halves (each 234/f feet long) insulated from each other by an insulator of approximately 2-4" length, and coaxial cable connected, shield to one side, center conductor to the other.

TABLE 7-1. Approximate half-wavelength antenna lengths.

FREQUENCY (MHz)	Half-wave resonant antenna approximate total length (feet)
3.6 MHz	130 feet
3.9 MHz	120 feet
7.1 MHz	65.9 feet
7.25 MHz	64.6 feet
14.15 MHz	33.1 feet
21.3 MHz	22 feet
29 MHz	16.1 feet

At HF frequencies the loss due to the coaxial cable (presuming an SWR of 2:1 or less) is not significant for transmission line lengths of less than 100 feet; almost any coax can be utilized. RG58, or RG8X are often chosen for powers below 500 watts. As discussed in the previous chapter, adding a simple toroid-based balun at the center may help reduce radiated RFI in the digital ham shack; a 1:1 current balun inside the shack (often called a radio isolator) may also help.

Almost any kind of wire can be used, but stranded wire will probably last longer. It can be insulated or not; either works. (Insulated may resonate with a slightly shorter length.) Depending on height of a horizontal antenna above ground (due to reflection and absorption by the ground), the impedance is 50-75 ohms, and typically 50 ohm coaxial cable (RG-58, RG-8 or similar) is utilized. The part of the radiation that proceeds downward bounces off the ground (at some depth) and creates an "interference pattern" with the upwardly radiating energy (with nulls and peaks).

The net result is that horizontal antennas that are roughly less than a quarter wavelength high tend to put more energy in the upward direction and are more useful for reaching nearby states (after bouncing back from the ionosphere). Antennas that are significantly more than a quarter wavelength high put more energy in the outward (lower angle) direction and are relatively more useful for distant contacts (because that peak at the lower angle will bounce off the ionosphere and come back to earth a thousand or more miles away).

Vertical antennas tend to put most of their radiation in low angles (with respect to the ground) and are frequently use for VHF/UHF line of sight transmissions, as well as for long-distance HF transmissions.

While the center connection to transmission line can be crimped or even twisted for emergency work, typically it is soldered. The end loop through the insulator may be soldered or even simply tied in a knot!

The antenna length is adjusted if need be, by using either an antenna analyzer or a simple SWR (standing wave ratio) meter, in an effort to get the SWR below 2:1 at frequencies on interest. Typical bandwidths of wire antennas are in the range of 2% of center frequency. Insulators can be made out of any non-conductive, non-brittle substance. Electrical PVC with holes drilled through works well; white PVC works but won't stand up to sunlight as long. There is no need to purchase expensive insulators.

In addition to the design frequency, the simple centerfed half-wave dipole may function adequately on odd, but not even harmonic frequencies. In order to cover more bands, additional dipoles may be connected to the same center conductor to make a "fan dipole." There will be some interaction between the dipoles; a bit of trial and error is generally required. Try to keep the wires from tangling; I typically separate them by a couple degrees of angle using wooden sticks at their ends.

A more complicated but more versatile HF antenna is a non-resonant antenna driven by an antenna tuner. The antenna tuner contains inductor(s) and capacitor(s) to allow impedance matching to a non-resonant antenna of widely varying impedance. Designs that are commercially available tend to use the T-network, Pi-network or L-network designs. Automated tuners are also available. If carefully given about 5-10 watts of power held steady on a frequency (e.g., by sending key-down CW or using the FLDIGI "tune" button), the automated tuner's microcomputer will run through a series of possible configurations of an L-network and find the best possible match.

A surprising innovation is that now at least one commercially available model actually measures forward and reflected values, and does computations to discover the likely match values, so that the search for best possible match is driven more by mathematics than by trying every possible choice. [52]

The automated tuners do not always succeed, but they certainly are easy to use. Typically they will also memorize the setting for that particular frequency. A useful length of non-resonant antenna is anything from 30 feet to 150 feet overall length. Traditionally, these are fed with balanced ladder-line transmission line, which is extremely low loss (unless iced) at high SWR's, and then either connected directly to the antenna tuner (optimally) or (less optimally) to a 4:1 balun allowing a modest length of hopefully large diameter lower-loss 50 ohm coax to go to the antenna tuner. The balanced line may be 300-ohm or 450-ohm, and stranded may last longer than solid. Traditionally one puts a few twists in the balanced transmission line to deal with asymmetries of nearby structures; keep balanced transmission line several inches away from anything metal if at all possible.

The standard automated L-network based tuners may have difficulty tuning at the frequency where the antenna is a full wavelength. Older 3-component adjustable pi-network tuners (or recent T-configuration tuners) seem to handle this situation more easily. A coax-trap set to this frequency that effectively shortens the antenna just for a fraction of a MHz around the resonant frequency of the coax trap works well to solve this problem, and coaxial traps are easy to make. [53]

In a real pinch, an L network antenna tuner can be made by coiling wire around PVC pipe with adjustable taps, and a variable capacitor can be created from aluminum foil sheets slid in and out of insulating manila envelope. Such a system is more likely to succeed at 80 meters than at 10 meters, however.

VHF ANTENNAS
The shorter wavelength of VHF/UHF frequencies makes it much easier to construct more elaborate antennas which make up for the decreased "aperture" of a smaller antenna. Typically vertical antenna constructions are utilized because communications are usually line-of-sight (so that the ionosphere does not change the wave's orientation) and mobile vehicular-mounted antennas are almost always vertically polarized. A vertical dipole can be built easily in a similar manner as the HF antennas described above and will have good performance.

1/4 wave vertical dipoles using a "ground" will have lower efficiency due to losses in the "ground" unless that "ground" is a substantial piece of metal -- like a car body. Multiple radials (1/4 wavelength) can be used if necessary. A magnetic mount 1/4 wave vertical intended to be used on a car body can be mounted on a large cookie tin or even better a filling cabinet for acceptable results.

In Gainesville, we have used omni-directional homebrew "slim jim" antennas which are an end-fed folded dipole, with a transmission-line based L-network matching system built in. These were built very inexpensively by using a pressure-treated 1x2 wooden strip as the support, and then 14-gauge regular house wiring (THHN solid conductor) as the antenna and matching network wiring. This got the cost well below $5.

The pressure treated wooden backbone is impervious to rain, and can be drilled to allow rope connection. We were able to easily hoist these high into trees using 1/4" paracord or nylon rope, and slingshots & fishing line to first gain access to a high limb. It is wise to also have a rope attached to the bottom of the antenna to allow it to be pulled down by the rope if it gets stuck, instead of having to pull the coax cable.

It's important to understand that height in this case has TWO advantages: 1) it dramatically reduces the losses due to ground-level absorbing obstructions, and 2) it reduces the take-off angle of the vertical antenna caused by the ground reflection. Thus in two ways, adding height will get you more distance.

In addition to being very inexpensive, these antennas have a natural EMP-hardening built in, because the shorted transmission line matching system basically eliminates frequencies significantly below the design frequency.

WOODEN SLIM JIM 2-METER ANTENNA CONSTRUCTION DETALS:

WIRE: Start with a piece of solid #14 AWG household wire approximately 3 yards and 9 inches long (117") (It is easier to be a couple inches too long and later nip the excess off.) Strip the insulation off of 36" at one end.. It is easiest to do this with a pocketknife while holding the wire against a solid flat surface.

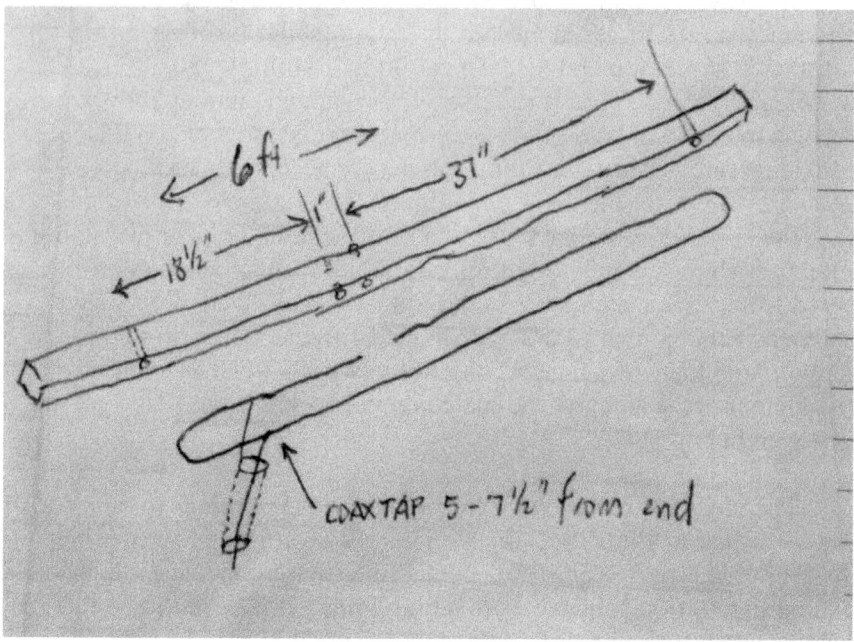

Fig. 7-1. Drawing of the wood and wire that make up the antenna. The end-fed folded dipole is the longer, righthand portion, while the transmission line matching network is the left hand portion of the wire to which the coax is attached. **Note: Making the matching section 19.5" instead of 18.5" sometimes makes this easier to tune.**

WOOD: Start with a pressure-treated 1x2 that is 8 feet long. These are typically less than $2 at home improvement stores. Leave several inches of space (perhaps 8") at one end to "hang" the antenna by, and drill a 1/4" inch hole through from front to back for later hanging. At 8" from the end (the "top" of the antenna) drill a 1/8" hole clear through from side to side. 37" from that first hole, drill a 1/8" hole just half an inch in to give you a stopping point for the folded dipole. Another 19.5"(up to 20.5") further down the wood, drill another 1/8" hole clear through for the shorting leg of the matching transmission line. The total length between the two through-and-through holes will then be 18.5" (to 19.5") (matching section) + 1" (gap) + 37" (folded end fed dipole) = 56.5 (or 57.5) inches, just a bit under under 6 feet).

Distances just aren't that critical. That 37" inch length is simply not critical. I tested 3" longer and 3" shorter and they still would work with just different coaxial cable attachment points, very little difference "shorter" and so with somewhat more difference with "longer". The matching transmission line distance probably isn't terribly critical, either, and increasing it to 19.5" has made better SWR's more easily attainable on several antennas.

CONSTRUCTION: Thread the wire through as needed to form the antenna as shown in the drawing, and secure it with electrical tape. Try to pull sections reasonably tight so the wire hugs the sides of the 1x2 wood. *Cut off any excess as needed* so that there is roughly a 1" gap between the free end of the matching line and the far end of the folded dipole end.

MATCHING: Using an antenna analyzer (with a very short connection, like 2-3" of wire) or an SWR meter (if possible, with a short connection, or a connection that is ½ wavelength (roughly 31" for RG8X) so the impedance isn't altered by your coax line), run the connection up and down the matching section. Use a finger on each side to make the connection, and keep them even with each other. You'll quickly find the point where you get an SWR very near to 1:1, often about 7" from the shorted end. Mark this, and solder the coax to it there, with the center conductor of the coax going to the longer side and the shield going to the side of the matching transformer that goes nowhere.

WATERPROOF: Use a liquid or grease sealant of your choice on the ends of the coax, run the coax either directly away from the matching loop or tape it right down the center. Secure the antenna wires and matching section every 12 inches or so with electrical tape.

FEEDLINE REFERENCE INFORMATION
Note that losses increase dramatically if the line has an SWR significantly greater than 1:1, and for ladder line, if the line is wet or covered with snow or ice.

TABLE 7-2. Transmission Line Loss Characteristics

Type of Transmission Line	100 foot loss in dB at 146MHz
RG58A/U (50 ohms)	6.1 dB

RG8X (50 ohms)	4.5 dB
RG 8 /LMR 400 (50 ohms)	1.5-2 dB (LMR is lower loss than RG8)
450 Ohm ladder line	0.4 dB

Remember! These loss values are when the transmission line is operated with a perfect match-- 1:1 SWR. When the SWR is higher, there are points with much higher voltages and much higher currents, resulting in significantly greater losses, particularly for COAX lines. Open wire feeders do much better with high SWR's, but are much more susceptible to losses due to rain, snow, or ice.

HOW IT ACTUALLY WORKS
(Rather technical information, that some readers might prefer to skip.)

The **Slim Jim** is a close brother to the **J Pole**. Both use a transmission line matching system. The only difference between the two is the Slim Jim bends the antenna back on itself to make a "folded dipole" half wave end fed, while the J Pole simply has a single wire end fed dipole. There is only a very small difference in their performance, not worth worrying about. And a center-fed dipole also has very similar performance! You could cut off the folded-back portion of a Slim Jim, re-tune as needed, and use it as J Pole. **A quarter-wave dipole needs a fantastic radial system or an incredible ground system to equal the performance of a half-wave dipole, slim J, or J pole**

The matching section is much more fascinating. Although often represented as a ¼ wavelength "transformer" moving geometrically from 0 ohms at the shorted end to near-infinity at the open end, it is actually a more mundane – and complicated.

The shorted end stub is actually acting like an **inductor** in parallel with the coaxial feed, part of an L-network (inductor-capacitor matching network). How the shorted section acts like an inductor is shown on the Smith Chart below.

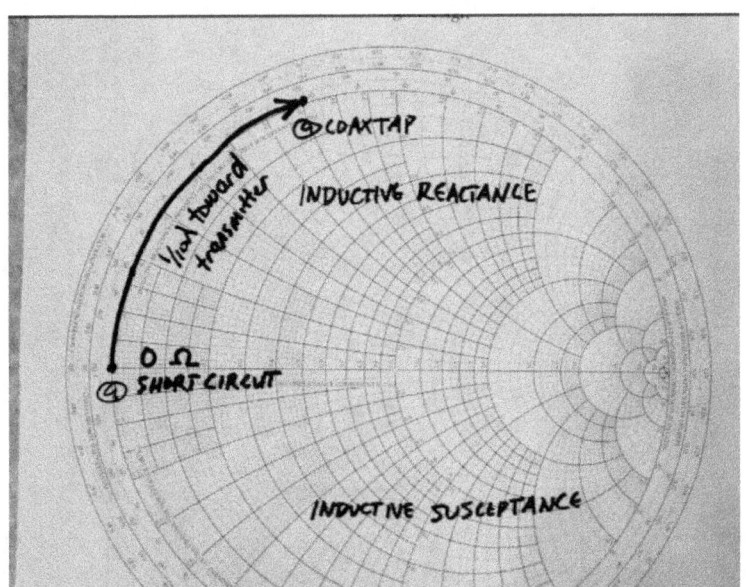

Fig. 7-2 Shorted section has 0 ohms at its shorted end, but 1/10 wavelength away it has considerable inductance, which is in parallel with the coaxial feed.

The apparently "open-ended" transmission line segment connecting the coax to the end-fed dipole probably plays two roles. EZNEC evaluation by others has shown that the currents in the two wires of the balanced line aren't exactly the same, and not exactly out of phase, either! So there is probably some antenna radiation from this "transmission line". In effect, it is providing a bit of antenna for the "other wire" of the antenna, making it not quite completely end-fed, but instead fed just INSIDE the end of the antenna (providing a lower impedance feed point).

The balanced feedline then performs matching duties by providing a series capacitance between the coax and the "near-end-fed" dipole. How the (almost) open-ended section acts like a capacitor is shown on the Smith Chart below.

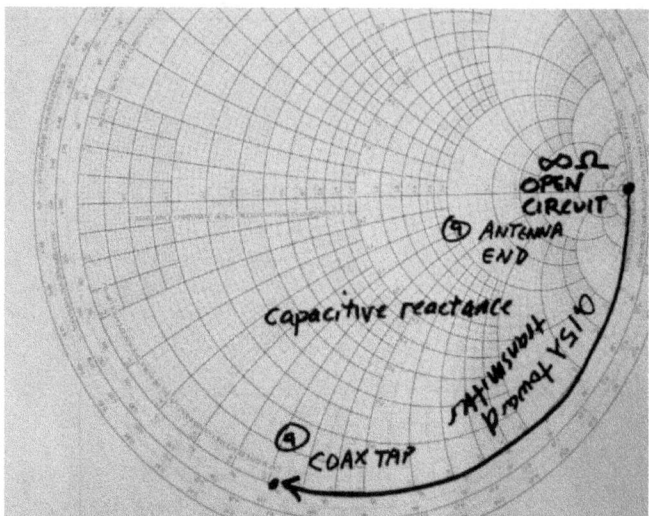

Fig 7-3 *Seemingly "open-ended" balanced line section has near infinite ohms (near open circuit, because the end impedance of a ½ wave dipole is very very high, and the second wire of the feedline goes NOWHERE) but 0.15 wavelengths back toward the transmitter (at the coax feed point) it has significant capacitance – which is effectively "in-line" (in series) with the antenna.*

The end result is that by picking the tap point with your coax, you are actually tuning the following L network that matches your coax to the antenna feedpoint impedance for best power transfer.

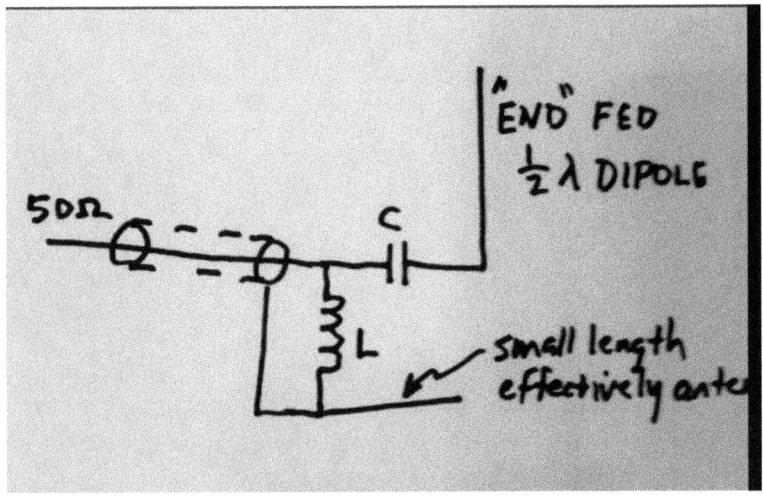

.Fig 7-4. *Equivalent matching schematic.*

8   BRIEF INTRODUCTION TO THE UNDERLYING DETAILS OF PACKET COMMUNICATIONS

This introduction to packet communications is by necessity, incomplete and may blur or gloss over distinctions that more erudite readers would make. However, it should give the novice reader enough information to begin understanding the cryptic information that packet systems often display in monitor screens! If your packet system is set up well by someone else, it will happily work quite well without your understanding any of this chapter. But if it malfunctions, understanding this chapter may greatly help you in figuring out what is going amiss.

The concept of splitting information to be transmitted down into discrete **packets** was developed in the 1960's. A packet includes **header information** that gives routing information, and a **payload** of the desired information to be transferred. Connectionless packet protocols such as Ethernet, Internet Protocol (IP) and User Datagram Protocol (UDP) require that the header include complete addressing information so that the packet is switched (possibly multiple times) in ways that will eventually get it to the desired destination. Connection-based protocols such as X.25 (and AX.25, the amateur radio version with which we deal here), Frame Relay, and Transmission Control Protocol (TCP) presume the prior effort of establishing the connection, so that the packet's header requirements are simplified[54] One type of packet may be embedded as information within another type of packet (example: TCP/IP)

The reader may better understand AX.25 Packet communications if they first observe the OSI (Open Systems Interconnection) Model layers. [55] This is a conceptual abstraction of networked communication systems into a series of layers. The bottom layer is the actual voltages, currents and wires (or else) that conveys the information; the top layer is the software application the user wishes to utilize. In between are additional layers, each building on the strengths of the layer below it and serving the needs of the layer above it, that accomplish the task of moving information reliably from one location to another. Many of the obscure acronyms (SOCKS, TCP, IP etc) that modern computer users are confronted with, come from software that functions at different layers of this model. The most common usage of networking – the Internet – does not perfectly conform to these well-defined layers and there are even arguments of how it maps to these layers.[56]

TABLE 8-1. OSI Network Layers

Layer	Name	Examples
7	Application layer	An email program; DNS; FTP; web browser, photo-editing program.
6	Presentation layer	MIME, jpeg
5	Session layer	NetBIOS, SOCKS
4	Transport layer	TCP, UDP
3	Network layer	IP (Internet Protocol); Appletalk
2	Data link layer	Ethernet frame, X.25, AX.25
1	Physical layer	RS-232, USB, 1200 baud Bell 202 tones, Ethernet packet

As an example an Ethernet packet (maximum 1522 octets) transmits across a physical medium (Layer 1), the job of which is to get the 1's and 0's transferred. Inside that Ethernet packet is a data "payload" which may be an Ethernet frame of up to 1500 octets. Each network interface connected to the network has a "burned in" unique Media Access Control number (from the manufacturer).

Layer 2 (Data link layer) has to get the packet to the right network interface, which are numbered at the factory with a unique Media Access Control (MAC) number. To get an Ethernet packet to the right Media Access Control (MAC) interface, the sending device may send out an Address Resolution Protocol request using a special "broadcast" number to which all devices listen. The reply will tell the sender the correct MAC number for the desired recipient. On a PC, you can see the cached table by typing the command **arp -a**.[57] Bad (erroneous) frames are detected at Layer 2.

Layer 3. Inside the Ethernet frame may be an IP packet (layer 3); the job of layer 3 is to transfer information from a specific IP number on a computer to a specific IP number on another computer. IPv4 was able to accommodate 4 billion different hosts; IPv6 erases that limitation. The Internet Protocol does not guarantee error free delivery.

Layer 4. Inside the IP packet (layer 3) may be a TCP packet (layer 4); the job of the 4th layer is to get the information to the right service (represented by a port number) of that IP number. (Common port numbers for many applications you may use on an everyday basis are listed in Table 10-2.) Errors can generate "send it again" requests at this layer. TCP at this level provides guaranteed error-free delivery. UDP is a simpler system that only provides "best effort" delivery.

At even higher layers, a portion of a .jpeg photo may be inserted into the payload that will go into the TCP packet, and a photo-editing program (Level 7) may choose the contents of the .jpeg file. [58] [59] [60]

GETTING BACK TO AMATEUR RADIO

AX.25 is roughly a 2nd layer protocol (much simpler than things that use all the layers introduced above). It attempts to provide error-free delivery of packets at this layer. It is not necessarily fixed to any one first (physical) layer. In Amateur radio, it has traditionally used 1200 baud Bell 202 tones (1200 Hz and 2200 Hz tones; the center of those is 1700 Hz).[61] Classic Terminal Node Controllers (TNC's) included both the tone-handling circuit and the layer 2 Data Link handling circuitry/software in the same device. This was done because computers of the day were not very powerful, so it was easier to use them as simple dumb terminals and let the TNC do all the heavy lifting.

Of course, as personal computers became so much more powerful, it became more reasonable to put more of the smarts in the computer and less in the expensive accessory device. In fact, as sound cards became commonplace in personal computers, the computer software evolved to not only handle the 2nd layer, but to use the soundcard for the tones of the 1st layer. While all that can be combined in one software application (e.g., MixW[62]), it can also be split between two applications.

The most typical split chosen involved an interconnection known as KISS (keep it simple, stupid!). Soundcard-based Bell 202 packet tone applications such as UZ7HO soundmodem.exe[63] (for Windows) or Direwolf[64] (for Linux) present a KISS interface, to which packet based software such as WINLINK RMS PACKET (for windows) or linbpq (for Linux) can connect.

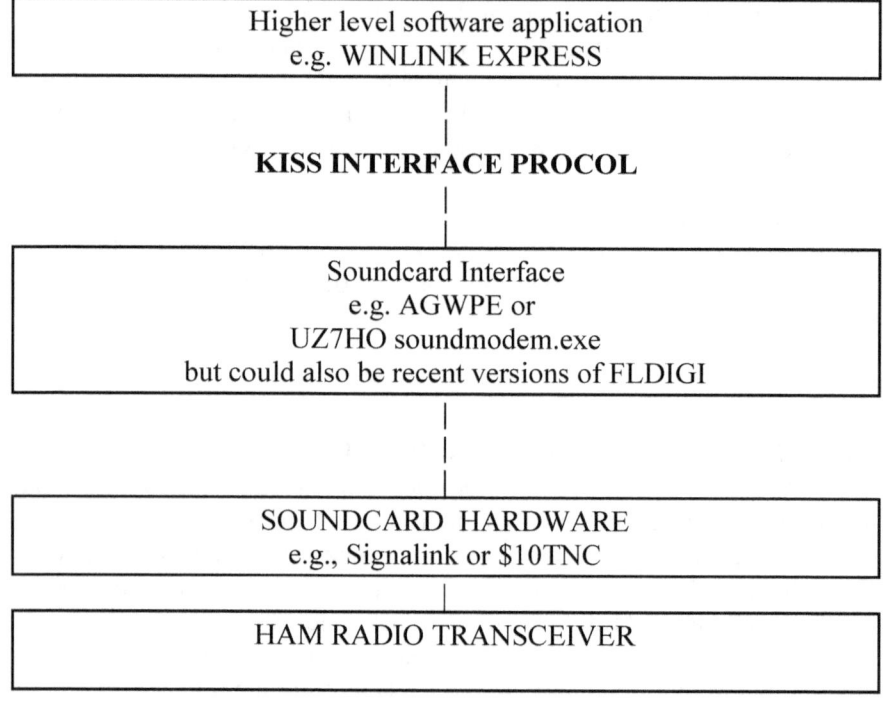

Fig 8-1. *Illustration of the position of the KISS Interface in the stack of hardware and software that makes digital work.*

An interesting fact, however, is that linbpq AX.25-based software can connect to any KISS modem software, not just Bell 202 tone-based software. Version 3.23 of FLDIGI provides a KISS interface,and can communicate by a dizzying variety of protocols such as PSK, Olivia, RTTY and others. If the underlying protocol includes the needed characters, AX.25 can happily be accomplished on any of several different protocols, no longer inseparably linking to Bell 202 tones!

AX.25 Version 2.0 protocol[65] allows for the transmission of three types of frames of information, each of which is made up of fields as follows:

U Frame or S Frame

Flag	Address	Control	FCS	Flag
01111110	112/560 bits	8 bits	16 bits	01111110

Information Frame

Flag	Address	Control	PID	Information Payload	FCS	Flag
01111110	112/560 bits	8 bits	8 bits	N * 8 bits	16 bits	01111110

Fig 8-2. *AX.25 Frames*

The **Flag field** serves as a start and stop delimiter of a frame.

The **Address field** indicates the source and destination, among other duties. (Ham radio callsigns)

The **Control field** describes the type of frame.

The **PID** (Protocol Identifier Field) allows for the transmission of various higher protocols, including TCP/IP packets, Appletalk, Internet Protocol and Net/ROM

The **Information field** carries the real message information, and may have a length up to 256 bytes.

FCS is the Frame Check Sequence, calculated by both sender and receiver and used to verify the accuracy of the transmission.

The meaning of the Control Field will often be made visible in monitor screens presented by amateur radio packet software. Observing that presentation can often clear up what is working correctly or incorrectly with a packet connection. The following are Control Field designators you may see often:

Table 8-2. Control Field contents

Meaning	Abbreviation
Receive Ready	RR
Receive Not Ready	RNR
Reject	REJ
SABM	Layer 2 connection request
DISC	Layer 2 disconnection request
I	Information Frame
UI	Un numbered frame
DM	Disconnect mode (system busy)

Information Frames are numbered; the numbers repeat consecutively from 0 through 7 and then back over and over.

Some important software setup parameters that you will often encounter and which relate directly to AX.25 transmission:

- A parameter MAXFRAMES is used to indicate the maximum number of information frames that can be sent without an acknowledgment of correct reception.

- A parameter MAXRETRIES defines the number of efforts to get a frame successfully across before giving up the connection.

- A parameter FRACK defines the amount of time a sender will wait for an acknowledgment of correct reception of a frame before simply sending it again.

TCP/IP & MESH In Relation To AX.25

Passing TCP/IP over AX.25

As you can see from the above, AX.25 is a lower-layer protocol, somewhat analogous to Ethernet frames. Just as TCP packets inserted into IP packets can be inserted into Ethernet frames, TCP/IP (or UDP/IP) can be inserted into AX.25 packets and some amateurs create tcp/ip network over AX.25. (This introduces significant overhead, however.) The Amateur Packet Radio Network (AMPRNet) is a world-wide example.

Passing AX.25 over TCP/IP

The reverse is needed when you have a working TCP/IP network (perhaps over a WIFI physical layer such as MESH systems) and you wish to forward AX.25 frames. These can be transmitted by inserting them into the data payload of a UDP packet (AX/UDP) for example , which then gets inserted into an IP packet, and so on to be transmitted by MESH. This is what the BPQAXIP functions of John Wiseman's software accomplish in order to allow node to node communications over TCP/IP-based MESH devices. Although UDP is not guaranteed delivery, the AX.25 portion will create retries as needed to get the message across.

9 BRIEF INTRODUCTION TO LINUX

Our group used the G8BPQ `linbpq` software running on Raspberry Pi's because with homebrew $10TNC sound-card interfaces, it was the absolute cheapest way we could get a large number of digital repeater stations created for backup and deployment. **If one cannot build sound-card based interfaces, then buying working or new Kantronics TNC's may be the best method, at around $200.** Using a $40 Raspberry, $5 memory, $15 backup battery, and $15 for the soundcard interfaces and cables, we were able to cut that portion of the total cost down to about $75 for each digital repeater, a significant savings for us. Used Windows-based laptops would be another alternative, but the cost would still rise significantly, probably because of the inherent cost of the Windows operating system license.

Using a Linux-based product forced us to gain some expertise with Linux. The purpose of this chapter is to give you a running start on Linux should you decide to build the Raspberry Pi digital repeater.

Linux is a variant operating system of Unix. It is available virtually for free for non-commercial usage. It is a multi-tasking multi-user operating system, which means that several different users can each be doing several different things simultaneously on your Raspberry Pi without difficulty. Linux keeps their programs completely separate and will not (normally) allow a crash or problem with what one person is doing to affect the others.

While Linux includes a huge amount of command-line capabilities it also includes a vibrant graphical interface that isn't that dissimilar to the familiar Windows desktop.

There is a well-developed user support system to help you get a system going from scratch. This basically involves taking a micro-SD card and creating initial software on it using a Windows-based computer and the Internet, and from there finishing the job on the Raspberry itself.

While you'll undoubtedly do quite a bit in the graphical interface of the Raspberry (because it works much like computers you're very used to), you would be wise to gain some understanding of the command-line interface, which will help you when you can't get something to work. A suitable tutorial (there are many) can be found at:
http://www.ee.surrey.ac.uk/Teaching/Unix/index.html

A few basic details here:

1. Linux uses directories (& sub-directories) just like Windows, except that the slant character in the written specification of a file location goes the other way.
2. Linux has many similar commands to Windows commands.
3. Linus has a more highly developed user structure, with users, and groups, and everyone, keeping separate track of who is allowed to read, write or execute a file.
4. On Linux, there are files and processes. Everything is either a file or a process.
5. Processes are numbered. The command `ps` will give you a list of running processes. `kill <number>` will allow you to kill a process (similar to using Task Manager to stop a process or application in Windows).
6. Linux includes the ability to have jobs (processes) run automatically at various times. `cron` is the facility that handles this.
7. Windows `mkdir` becomes Linux `md`
8. `ls` will allow you to see what is in a directory
9. `ls -l` will add more details to what is in a directory
10. `.` refers to the current working directory
11. `..` refers to the parent of the current working directory
12. There are hidden files in Linux, just as in Windows.
13. Different flavors of Unix store important files in different subdirectories. `/var/logs` is where log files will build up on

your Raspberry Pi and it would be wise to check the size of them periodically and perhaps erase them when you need more space.

14. Linux can run processes both in the foreground (visible) and in the background (invisible), just like Windows does.

HELPFUL LINUX COMMANDS

ls -l	Gets detailed directory listing
cat <filename>	Print out a file
cat <filename> >more	Print out file page at a time
cd <directory>	Change "current directory" to <directory>
ps -e	Show all processes running
passwd	change password
pgrep <program>	Find process number of <program> if running
kill <process number>	kill a process (use carefully)
man <command>	Request information on a command

10   RASPBERRY PI-BASED DIGITAL REPEATERS

OFFER OF ASSISTANCE
Configuring a Raspberry Pi with all this software is quite complicated, and takes hours.   However, once you have a working system, you can use onboard diskcopy utilities (in the graphical menu!) to easily make a clone of your system.

In order to assist other groups developing the G8BPQ `linbpq` system, until it becomes overly burdensome, I'm offering to make you a complete system on a chip with your callsign if you follow the following directions carefully:

1.   Send me an email at docvacuumtubes@gmail.com (if that doesn't still work, look up my current email address (KX4Z) on qrz.com) to find out if I'm still able to offer this free service.
2.   Mail me a 16 Gbyte micro-SD chip (to the address that I return to you via email) packaged securely and with a self-addressed envelop with correct postage to get it back to you. I suggest adequate cardboard to protect the micro-SD chip, and likely inserting it inside its protective SD card adapter (usually comes with it when you purchase them.)
3.   Your amateur radio callsign, 6 character or less maximum, and 6-character maximum NODEALIAS.
4.   Whether you will be using a TNC-X terminal node controller (NOT TNC-PI), or an Adafruit Audio Adapter-based soundcard

system, MFJ-1204 or other "classless" sound dongle, or a Tigertronic Signalink. I need to know this to properly configure `linbpq`, and `direwolf` if needed. Note that because `alsamixer` doesn't interface properly to the Signalink at this time, I may not be able to guarantee correct volume levels with a Signalink, though you can adjust the front panel controls.

5. I will copy all the software that works on a Raspberry Pi version 3, including the operating system, and insert your callsign (SSID-7). The options for chat server and WINLINK CMS server (in `/home/pi/bpq32.cfg`) will be commented out, but present so that if you learn how to enable them, you can easily.

6. I will drop it back into the United States mail and hope that it makes it back to you.

7. (Unfortunately) this offer is without any guarantees of any kind. You agree to hold me harmless for any problems the chip causes.

RATIONALE

Digital VHF emergency communications that need to go farther than possible by simplex communications will require a digital relay station, typically known as a "node". Nodes typically offer some or all of the following services:

- Allow connections, including sequential connections, so that you can reach further nodes (packets are error checked and re-transmitted as needed, a more efficient error-free transmission system than digipeating)
- Maintain a list of routes to various distant nodes
- Simplify connection to distant nodes
- Provide digipeating (packets are passed through without error checking)
- Chat room services
- May offer bulletin board services (a rudimentary email type system)
- May offer connection to WINLINK email over radio system

Although this chapter discusses one specific method of creating a digital node station (`linbpq` on a Raspberry Pi), there are many other alternatives, including:

- Purchase a Kantronics KPC-3+ hardware-based TNC[xii]
- Utilize BPQ32 on Windows [the windows version of BPQ software--make a full node on a Windows computer]
- Purchase a working used TNC made by any of several vendors.

Because I purchased several used TNC's only to have significant difficulties getting any of them to work, and new Kantronics KPC-3's are a moderately expensive item, and I did not want to dedicate a Windows computer to this purpose..... I leaned how to install a Raspberry Pi linbpq system. Although this chapter is primarily addressed toward VHF server stations, the same concepts work on HF.

Purpose

These instructions will help you create a Raspberry Pi- based Linux BPQ (linbpq) "node". The developer of the BPQ series of software is John Wiseman G8BPQ, who is larger than life in packet circles. His web site includes an enormous amount of documentation,with which you will become very familiar.

linbpq software will require some form of terminal node controller to connect to a radio.   Several TNC options,with their advantages and disadvantages are shown in the following table:

TABLE 10-1.   Methods of connecting to a radio transceiver

No.	Description	Advantage	Disadvantage
1	MFJ TNC-X or similar KISS-compatible TNC[66]	Hardware solution works well.	Price
2	Direwolf linux software + alsamixer-compatible USB soundcard (e.g. Adafruit audio adapter) + Slignalink SL-1+ or similar	Adafruit audio adapter only $5.	Signalink SL-1+ is only available *used* (try Ebay)
3	Direwolf linux softwre + alsamixer-incompatible Signalink-USB	Plug and play solution.	Difficulty predicting sound levels

xii This is widely available. For example, http://www.universal-radio.com/catalog/tnc/1908.html or http://www.hamradio.com/detail.cfm?pid=H0-000229

			due to incompatibility (so far) with alsamixer
4.	Direwolf linux software + alsamixer-comptabible USB soundcard (e.g, Adafruit auido adapter) + homebrew version of the Signalink SL-1+ ("$10 TNC")	By far the cheapest solution.	Requires about 2 hours of construction time for an experienced audio circuit builder.

I have personal successful experience with options 1, 3, and 4 and have no doubt that #2 will work very well. Since our group has built >10 homebrew "$10TNC's" we have primarily gone the route of #4. However, if building circuits isn't your forte, simply either going the route of a Kantronics KPC-3 (no Raspberry needed!) may be a better choice, or else using option 1, with a ready-made MFJ-TNC-X.

Purchase List

Raspberry Pi Computer & Related
Suggested examples of commercially available items in 2016.

1. Raspberry PI Version 3 (which means it is a "B") $36[67]
2. Clear protective plastic case with heatsinks: $6 [68]
3. Micro SDHC memory (and some Windows-based computer that will allow you to do the initial formatting and loading of that card). Class 10 means fastest. 8Gbyte minimum; 16 Gbyte may be a better choice.
4. Recommended: Large cellphone "backup" battery that is able to be charged and used *at the same time*, and has the micro-usb connector to provide power to the Raspberry PI. This allows you to create the effect of a UPS-- uninterruptible power supply--for your Raspberry PI. While not an absolute necessity, this avoids difficulties if your power ever glitches.[69]
5. High capacity USB cell phone charger with the micro-B connector used for Android phones. Recommend >= 2 Amps
6. Keyboard/mouse: I recommend that you utilize a Logitech

MK270 Wireless Keyboard/Mouse combo, which reduces your vulnerability to radio frequency interference. This particular product has been tested to work correctly even when the USB bus is slowed to USB1.1 speed to work with the Adafruit USB audio adapter: There are others that also work acceptably at the USB1.1 speed, but not all do.[70]

7. HDMI monitor, or a VGA connector-based monitor with converter to HDMI.

Note: In all the following, when you need to type something into a terminal window from your Raspberry Pi, I use `Courier TypeFont` **for exactly what you need to type.**

Format your SDHC card, using a Windows based computer.

- Purchase an SDHC card that is between 8 and 32 Gigabytes (8 is plenty but 16 Gbyte might be useful).
- Using an appropriate adapter if necessary, install the new micro SDHC card into your Windows computer and CAREFULLY NOTE WHAT DRIVE IT BECOMES. (You don't want to accidentally erase your C: drive on your computer, right?) Use the SD Association's Formatting tool.[xiii] Download this tool from *sdcard.org*. Set "FORMAT SIZE ADJUSTMENT" option ON in the "options" menu to make sure the entire card id formatted.

Note: a helpful expert has published an alternative way to configure a Raspberry with linbpq here:
http://www.prinmath.com/ham/bpqHOWTO.htm

PreLoad your SDHC card with NOOBS

1. Stands for "New Out Of the Box Software." Allows you to get

---

xiii https://www.sdcard.org/downloads/formatter_4/eula_windows/index.html

your Raspberry operating system of choice.

2.   On your Windows computer with the SDHC card still installed, navigate to: www.raspberrypi.org, click on DOWNLOADS button. Download NOOBS as a ZIP file. I suggest using "offline and network install." The version as of 2016-05-27 is 1.9.2, 1.02 gigabytes. *That might take a while, depending on your network connection speed.* I don't operate an Apple computer but it probably works similarly for this and the next step.

3.   Unpack the ZIP file onto the SD card that you're going to use. (On the computer I use, when you click the file from within the Windows Explorer it offers you the option to Extract All and to specify where you want it to go.) Once again, be CAREFUL exactly where you extract this to -- the extraction is large, 91 files comprising about a gigabyte.

Bring Up Your Rapberry PI Operating System

1.   Plug the SDHC card into your Raspberry. Connect up a mouse and keyboard to the Raspberry, using the USB slots farthest from the Ethernet connector (nearest the edge of the card) and an HDMI monitor. Get ready, but don't yet connect up the micro-usb power to the card. Follow the directions on this video to get your Raspberry started. https://www.youtube.com/watch?v=y4GOG4P-4tY   Although you may be offered a chance to connect to your WIFI system, it isn't necessary because you downloaded the NOOBS version that included "offline" installation of RASPBIAN.

2.   When offered, click the box by the RASPBIAN operating system and then click INSTALL.   This is going to take a while, as it is going to write 3209 Mbytes worth....on my system at 5 Mb/second, so about 10 minutes worth.

Optional: Additional video with more ideas for you: https://www.youtube.com/watch?v=-6OGuhLtKbU

3.   *Note: On my system, installation automatically set "boot directly to desktop" and "desktop log in as user "pi", so that the system would nicely turn itself on.   If you are offered options related to this, be sure to set them accordingly.*

4.   Once Raspbian comes up, you can go into the
MENU | Preferences | Raspberry Pi Configuration

and set the nation, timezone, and country for WIFI -- after which my WIFI began to work properly. You'll probably need to put in your WIFI system's password etc. You need to be on the Internet for the next portion.

5.   Under the tab "Interfaces" I enabled the **I2C** and **Remote GPIO**

Update and Add Sound Library to Your Raspberry Pi

Bring up a terminal window (the icon on the menu bar that looks like a black chalkboard). Your prompt starts out with a $ because you aren't superuser. If you type the command "`whoami`" it will tell you that you are user "pi"

Typing the command "`date`" will give you the date and time. "`info date`" will give you the syntax if you need to fix anything. I fixed the time with "`date hhmm`" After that my updates went better.

```
UNIX HINT:   to kill any terminal screen (or the thing running in
it) type CTRL-C
```

Updating your onboard software to the latest fixes:
```
sudo apt-get update
```
        (I got some errors on this one)
```
sudo apt-get dist-upgrade
sudo rpi-update
```

Install libasound2-dev package
```
sudo apt-get install libasound2-dev
```

Install a graphical user interface program scheduler
```
sudo apt-get install gnome-schedule
```

Download DireWolf C-Source Code and Make Your Own Executable
Note: if you're going to use a TNC-X, you can skip Direwolf; you won't need it.

Using the built-in browser on your raspberry (looks like a globe) download the latest unix DireWolf sourcecode. This can be done from https://github.com/wb2osz/direwolf/releases

Click on and download **direwolf-1.3.tar.gz** (or whatever the latest appears to be) and use that new name in the all the following if it has been updated beyond version 1.3.

My downloads showed up in the directory
/home/pi/Downloads

I then made a directory for them
```
mkdir   /home/pi/direwolf
```

Copy what you downloaded to that directory:
```
cp /home/pi/Downloads/* /home/pi/direwolf
```
Move yourself to that directory
```
cd /home/pi/direwolf
```

And extract all the files from the "tarball"
```
tar -zxvf direwolf-1.3.tar.gz
```

Note for your education, those options include:
-z work on gzip compression automatically
-x extract archives
-v verbose output so you can tell what is going on
-f read the following file as input

The extraction process will cause a new subdirectory **direwolf-1.3 to be** created and populated with all the files. Move there:
```
cd /home/pi/direwolf-1.3
```

Compile the entire source code and make an executable on your own raspberry pi (this is amazing): [replace the version with whatever version you obtain]

```
make -f Makefile.linux
```
This will take a while and compile a BUNCH of c code.
```
sudo make install
```

(This what the program recommend and what I did; as root)

```
make install-conf
```
Gives you an initial configuration file
```
make install-rpi
```
Gives you a desktop icon
```
cd /home/pi
```
and you can find the conf file there
```
ls
```

Dire Wolf is now installed and you should see an icon on your desktop for it.

Configure Your New DireWolf Sofware-TNC

If you haven't done so already, plug in your USB-based soundcard (which might be an altered sound dongle) into the slot that is high and nearest to the Ethernet 10BaseT socket. To be sure this will work, reboot your Raspberry.

Next find your sound card's device address with this command:
```
aplay -l
```

If everything is working you'll see the default (Raspberry hardware) ALSA audio device at card 0 and your USB TNC (whichever one you chose) at card 1: (amongst the gibberish)
```
card 0:   ALSA..........device 0
card 0:   ALSA....,,,,,,,device 0
```
card 1:   **[something related to USB or CODEC] device 0: [something related to usb]**

Remember that you're on card 1.

Now hunt for your mic input on your TNC
```
arecord -l
```

The Raspberry itself doesn't come with an *input* so you should see only
```
card 1:     [something related to USB or CODEC for a Signalink]:
device 0  [more mention of USB]
```

Armed with this information we can now use any Text Editor to edit the **direwolf.conf** configuration file. A Text Editor doesn't put in extraneous formatting stuff the way a wordprocessor does.... There are several possibilities on your Raspberry Pi:

> **Menu | Accessories | Text Editor** PREFERRED OPTION
> nano   A small text editor
> vi   A really complicated text editor

Unless you are a Linux Guru I suggest you use the (graphical) Text Editor because it looks like something you're likely familiar with and it works. Just for reference sake, I include the following brief information on the ubiquitous `vi` editor (don't use this unless you're forced to, it can BITE):

Brief intro to `vi` commands

> `vi filename`   opens filename for text editing
> vi has two modes, <u>command</u> and <u>text.</u>
> `[ESC]` puts you back into <u>command mode</u> -- do this whenever you are confused. arrow keys work to move you around in command mode. *Don't ever go into text mode until you are sure you are exactly where you want to be*
> `dd`   delete current line
> `i`   puts you into <u>text mode</u> and will begin inserting IMMEDIATELY
> `a`   puts you into <u>text mode </u>will begin appending IMMEDIATELLY
> **DO NOT USE THE ARROW KEYS in text mode.**
> As soon as you are finished editing, hit `[ESC]` and get back into command mode, otherwise you risk adding the darndest text in the weirdest locations you ever imagined....which can have disastrous results.
> `:q`   quit
> `:w`   write out to the filename
> `:q!`   quit without writing
> `/text` ` find text in the file

Are you sufficiently convinced this is not for beginners?

To be the safest, *__use the built-in text editor__* -- go to **Menu | Accessories | Text Editor** and use it to find your direwolf.conf file and edit as follows, then SAVE when you're done.

Open /home/pi/direwolf.conf. In the Text Editor you'll have to click on a few directory-structures, but with a bit of trying, you'll find it.

1.   In the FIRST AUDIO DEVICE PROPERTIES you need to have an ADEVICE statement that fits with whatever soundcard system you're using. There are several options, pick and try until you get one that works with your setup.

ADEVICE plughw: 1, 0
> Used to work with both Signalink and USB soundcard dongle, but seems to fail now, possibly due to updates to firmware in my Raspberry

ADEVICE plughw: CARD-Set,Dev=0
> Works with USB Adafruit audio adapter

ADEVICE plughw:CARD=CODEC,DEV=0
> Works with Signalink

2. In the section CHANNEL 0 PROPERTIES edit the MYCALL to replace NOCALL with your callsign -- and probably with a -SSID, usually 7, such as

    MYCALL   K1ABC-7

3. Leave the MODEM 1200 line uncommented.

4. Go down and verify that you have

    AGWPORT 8000
    KISSPORT 8001

5. You can create a beacon to allow you to test that this is working and direwolf can transmit:

```
PBEACON delay=1 every=2 overlay=S
symbol="digi" lat=00^00.00N long=000^00.00W
power=5 height=5 gain=0 comment="[your call]
Test Direwolf"
```

> ***Don't forget to turn this off later on after you have it all working***

6. Now save the file with the original file name (direwolf.conf)

Testing your homebrew or sound-card based TNC

Plug in your radio to your TNC and your TNC into the USB slot. Tune to some frequency where you can receive packet signals, click on Direwolf icon to get it going and see if it can hear packet signals. Adjust the volume so that it works, and so that the volume number is around 50.

Adjust the transmitter potentiometer on your TNC so that your signal is just below maximum amplitude (an attempt to set the deviation within proper limits).

Once you have this working, you probably want to go back into the direwolf.conf and comment out (#) the PBEACON line because you're going to have your NODE do the beacons, not your interface software, from now on.

Setting up the LinBPQ Node

Instructions direct from the author of the BPQ software:
http://www.cantab.net/users/john.wiseman/Documents/InstallingLINBPQ.htm

Other instruction sources that were very helpful to me:
http://www.wcares.org/special-interests-3/aprs/aprs-raspberry-pi-virtual-tnc/
http://www.wcares.org/special-interests-3/aprs/aprs-raspberry-pi-virtual-tnc/
https://philcrump.co.uk/Raspberry_Pi_APRS_Digipeater

http://www.seapac.org/documents/pdf/2015-n7qnm-Implememting%20an%20APRS%20Digipeater%20on%20a%20Raspberry%20Pi.pdf

An already-built pi version of LinBPQ (bpq software for linux) named plinbpq is available at:

http://www.cantab.net/users/john.wiseman/Downloads/Beta/

> *NOTE: the "dropbox" location specified in Version 1 of this book is no longer operative, and a change in the WINLINK reporting protocol has rendered that older version of plinbpq no longer workable for WINLINK purposes; hence the new address above for downloading plinbp. Wiseman keeps his latest versions accessible via:*
> *http://www.cantab.net/users/john.wiseman/Documents/Downloads.*
> *html which should be investigated for more recent upgrades, if you are reading this book significantly after June, 2017.*

Create a subdirectory to hold the software and configuration files:
```
mkdir /home/pi/linbpq
cd /home/pi/linbpq
```

Download John Wiseman's latest version (this wouldn't fit on one line but it is really one command):

```
wget
http://www.cantab.net/users/john.wiseman/Down
loads/Beta/pilinbpq
```

Rename the program to linbpq, then make it executable
```
mv pilinbpq linbpq
chmod +x linbpq
```

You also need some web pages for the management interface. Create directory HTML (capitals) under your linbpq directory, and download and unzip
https://http://www.cantab.net/users/john.wiseman/Downloads/Beta/HTMLPages.zip into it.

```
mkdir /home/pi/linbpq/HTML
cd /home/pi/linbpq/HTML
wget
```

http://www.cantab.net/users/john.wiseman/Downloads/Beta/HTMLPages.zip

(all on one line; the line break here was introduced by the word processor)

```
unzip HTMLPages.zip
cd ..
```

Configuration of your linBPQ Node

The executable does not come with the required `bpq32.cfg` configuration file. This file is very important!

There are now (at least) *two ways* to create your bpg32.cfg file:

1. Willem Schreuder, AC0KQ, has written a web page that helps you automatically configure a bpq configuration file -- and even load linbpq -- http://www.prinmath.com/ham/howto/quickstart/

2. Here are links from my website with sample files you can edit:

    *   Simple html text: http://qsl.net/kx4z/bpq32.html
    *   PDF: http://qsl.net/kx4z/bpq32.pdf

    1. Using your browser on your Raspberry Pi, load the sample file into a window.   The select the entire text and paste it into the Text Editor.   Then save it in the same directory as your linbpq (/home/pi/linbpq)   Edit the radio port lines as appropriate depending on whether you are using a $10 TNC, a Signalink, or a TNC-X (the latter of which is the only one that doesn't need direwolf, and would connect to USB0)

    2. Edit to put in your callsign (with desired -SSID's) everywhere you see a call sign.

NO MATTER HOW YOU RETRIEVE THE TEXT OF THE FILE, WHENYOU SAVE IT IN YOUR /home/pi/linbpq DIRECTORY t he file **has** to be named exactly **bpq32.cfg**   The case of the letters is

important! Further information on this all-important configuration file is available here:
http://www.cantab.net/users/john.wiseman/Documents/BPQCFGFile.html

Before you turn on your system for the first time with a real radio connected, be sure you have adequately hardened your setup against RFI (Radio Frequency Interference) ....which can cause some really strange things to happen software freezing, transmitters locking ON, etc.

RADIO FREQUENCY INTERFERENCE
[Also see the discussion of this in the Client Station chapter.]

It is important that you deal with radio frequency interference to avoid possible damage to your setup. There is so much RFI near even a 5 watt walkie talkie that I have seen analog voltmeters "peg" on the 10 volt DC scale connected to NOTHING. In some ways, a handi-talkie is the WORST offender for RFI. Why? Where is the bottom half of that rubber-duckie dipole? It is the radio itself -- including any wire connected to it.....which would include the audio wires going to your TNC or soundcard!!! Since you aren't HOLDING the walkie talkie when it is used on a Raspberry Pi, the full RF current of the "other half" of the dipole can head towards your precious electronics!

Here's how to protect your systems: (see also similar information in an earlier chapter)
- Put ferrite clamp-on beads on the cables from the radio to the sound card, and if 1possible, between the sound card and the Raspberry (that lead is short).
- Install two or three 2" loops in the cable from the radio. Use electrical tape or a zip tie to hold them in place.
- Notice that the TNC-X is in a shielded box. So is the signalink. Your Raspberry is NOT, and if you used the $10 TNC, it isn't either! Most of my rigs work fine without any shielding around the Raspberry, but if you have trouble, put insulating baggies or other material (such as an envelope) around the $10 TNC and put the Raspberry PI in a plastic case (if you haven't already). Now put the entire setup inside aluminum foil (you can even just wrap the system or put some on the outside of a gallon baggie that makes it easy to move them in and out. Use some

aluminum foil to make a "shield braid" for the cable heading to the radio, and connect it to the aluminum that shielding the Raspberry and the $10 TNC.

- Get everything running before you turn on the radio. Then remove the USB cables for the MOUSE and KEYBOARD (they aren't needed once the application is started), get everything shielded, and THEN turn on the radio.
- If you have to back out of the software or fix something, turn the radio OFF first, and then plug the mouse and keyboard back in and go about your repair work..

Running your Node For The First Time

Finally! We get to run the thing. If you are using a sound-card type TNC, first double click on the DireWolf icon and it should start and look for a connection. You'll see an announcement in its window when it gets a connection from linbpq.

Then startup a terminal window and get into your linbpq directory:

```
cd /home/pi/linbpq
```

Start the program

```
./linbpq/linbpq
```

(note the period in front)

In a later version of this document I'll include information how to get the file to automatically start -- and stay running!

Install TELNET application on your Raspberry
Bring up a terminal window.
```
sudo apt-get install telnet
```

Now once your linbpq is running, you can connect to it using telnet, in addition to a packet over-the-air connection. (If you know your IP number [ifconfig will tell you that , you can even connect from

TELNET on another computer. Telnet no longer automatically comes on Windows machines, but you can re-enable it; you can also download PuTTY, which offers both telnet and more secure ssh connections.)

On your own raspberry Pi, it is easy:

```
telnet localhost 8010
```

Being able to telnet into your own linbpq application is a skill that should become second nature to you....

(The 3rd port in the supplied config file accepts telnet connections.)
In order for this to work, you will need to have included a user name and password for yourself in the TELNET port section of `bpq32.cfg`

Here's a screen grab of using PuTTY to log into my own server and do some maintenance checking: (my typing in bold)

```
--------------------------------------------------------------------------
login as: pi
pi@192.168.1.200's password:

The programs included with the Debian GNU/Linux
system are free software;
the exact distribution terms for each program are
described in the
individual files in /usr/share/doc/*/copyright.

Debian  GNU/Linux  comes  with  ABSOLUTELY  NO
WARRANTY, to the extent
permitted by applicable law.
You have new mail.
Last login: Wed Dec 28 18:00:13 2016
pi@raspberrypi:~ $ telnet localhost 8010
Trying ::1...
Trying 127.0.0.1...
Connected to localhost.
```

```
Escape character is '^]'.
user:gordon
password:

Welcome to KX4Z Telnet Server
 Enter ? for list of commands


?
JNVILL:KX4Z-7} CHAT RMS RELAY CONNECT BYE INFO
NODES PORTS ROUTES USERS MHEARD
mh 4
JNVILL:KX4Z-7} Heard List for Port 4
W4DFU-7     00:00:00:04
NK3F-7      00:00:05:00
NK3F-4      00:00:16:55
nodes
JNVILL:KX4Z-7} Nodes
AMHAM:KI4QBZ-7                       ARTGNV:KM4YGH-7
GARC:W4DFU-7          GNVRLY:KX4Z-11
GNVWLK:KX4Z-10                   LKCTY:KE4BQI-7
ORL:KF4LTF-6         POLK:WC4PEM-7
TOMGNV:NK3F-7
C TOMGNV
JNVILL:KX4Z-7} Connected to TOMGNV:NK3F-7
mh 4
TOMGNV:NK3F-7} Heard List for Port 4
KX4Z-7      00:00:00:00
NK3F-7      00:00:00:06
W4DFU-7     00:00:00:31
WC4PEM-7    00:00:00:50
KM4YGH-7    00:00:01:29
KI4QBZ-7    00:00:05:31
KE4BQI-7    00:00:07:03
KX4EOC-7    00:00:11:14
KI4KYS-4    00:00:13:50
KX4Z-4      00:00:17:15
NK3F-4      00:00:17:22
KI4QBZ-4    00:00:17:23
KI4KEA-7    00:00:17:53
KF4LTF-6    00:00:33:06
KM4YGH-4    00:00:44:55
KI4KEA-4    00:00:44:58
```

- -

Fig. 10-1   *Example transcript of using PuTTY to log into my own Raspberry Pi (using SSH), connecting into the* linbpq *software, and making radio connections.*

Monitor your system with a web browser (alternative to telnet)
Using your raspberry's browser, navigate to http://localhost:8080, or using another computer's browser to http://[your ip number]:8080 and guess what, you have web access to control your Node software!

Here's an example of what that looks like, where I have selected the Nodes choice to see which nodes my home server knows how to reach:

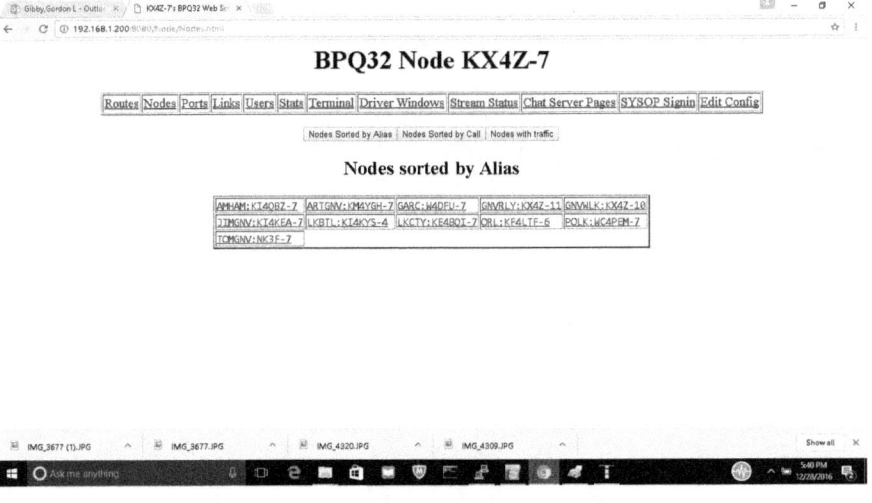

Fig 10-2.   *Web-based management of the* linbpq *system.   Port 8080.*

You can even get a terminal screen access, and edit your all-important configuration file bpq32.cfg using the web access.

Making your software start up automatically and re-start if it crashes

We take advantage of the linux **cron** daemon to get this to happen for us.

First we position two *scripts* within /home/pi. One script checks on direwolf, and if not running, starts it. The other does the same for linbpq. You can capture the verbiage for these files on the web, paste into the Text Editor and save them, or you can type them in yourself:

filename: /home/pi/direwolfscript
Web: http://qsl.net/kx4z/direwolfscript.html

```
# find direwolf process identifier
/usr/bin/pgrep direwolf
# if not running...
if [ $? -ne 0 ];
then /home/pi/direwolf-1.3/direwolf
fi
```

filename: /home/pi/linbpqscript
Web: http://qsl.net/kx4z/linbpqscript

```
# find linbpq process identifier
/usr/bin/pgrep linbpq
# if not running...
if [ $? -ne 0 ];
then /home/pi/linbpq/linbpq
fi
```

Next we have to get these scripts called every 2 minutes from cron. This can be done by manually filling out the crontab file by issuing the command

```
crontab -e
```

which will first ask you for your choice of editor (select nano) and then allow you to edit the file, write it out, and close.

Or, since you installed the graphical Task Scheduler, you can go to the Menu:
Menu | System Tools | Scheduled Tasks

and enter in two new tasks:

== ======= FIRST TASK =============
Command   **/bin/bash /home/pi/direwolfscript >/dev/null**
Default behaviour
Advanced: (click)
Minute: */2
Hour   *
Day   *
Month   *
Weekday   *

(Apply)
========== = ===================

== ======= SECOND TASK =============
Command   **/bin/bash /home/pi/linbpqscript >/dev/null**
Default behaviour
Advanced: (click)
Minute: */2
Hour   *
Day   *
Month   *
Weekday   *

(Apply)

========== = ===================

Making your Raspberry Reboot Periodically

There is the possibility you may have some RFI to your USB-port connection to a Signalink, or other interface TNC. In my experience, this will shut down the port, and Direwolf will be unable to function, hence your system quits performing. (**plughw** fails) In my limited experience, one system in a low-RF environment appears to have zero problem with this, while another, in a higher-RF environment has locked up the USB port occasionally.

The problem is that unless you provide for this, it *remains* down.

There may be another solution, <u>but rebooting the raspberry pi does</u>

work. Hence you may wish to add a cron job to reboot your raspberry pi at intervals. Rebooting is very quick, and the unit takes only a minute to do so. In my case, I selected every six hours. Unfortunately, user "pi" can't do this -- the cron job will fail because the operating system will ask for a password authentication. The way around this is to have root set up a cron job to reboot.

I don't know exactly how to do this in the graphical system, but it is reasonably easy to do in a terminal window. So start one up, then

```
sudo su
```

to grab root authority, then

```
crontab -e
```

to open an editor to edit the `crontab` file for root (which is different from the one you effectively created above for user pi, to start `direwolf` and `linbpq`.

Then use editor nano (or vim if you prefer) to put in the following line:

```
0 */6 * * * /sbin/shutdown -r now >/dev/null
```

the second entry is */6 which means midnight, 0600, 1200, 1800 (hours divisble by 6); there are five total entries (minute, hour, day, month, day-of-week) and spaces between them.

Another housekeeping item you might want to take care of within your root crontab table is to delete various chatconfig.conf.bakX and other .bak files that accumulate.... you can create a script that looks something like this:

```
rm /home/pi/chatconfig.conf.bak*  >/dev/null
sleep 4
```

and then call that script just before the shutdown/reboot command, with another line in the root crontab table (see below) The purpose of the sleep commands is to let the system finish writing out the changes to non-volatile memory

```
    0 */6 * * * /bin/bash /home/pi/cleanupscript
>/dev/null
    2 */6 * * * /sbin/shutdown -r now >/dev/null
```

Setting the Audio Gain (Volume) Levels

You'll need to set the mic gain and audio output levels if you're using a soundcard-based technology. If the audio output isn't at 100%, Signalink and similar devices (such as the $10 TNC) may not key the transmitter properly.

```
alsamixer
```

will bring up a "somewhat graphical" volume control. Use Left/Right cursor keys to reach your sound-card device. Use the up/down arrow keys to adjust volume. You probably want the audio output at maximum and the mic gain at 50%.

```
sudo alsactl store
```

SETTING YOUR USB PORT SPEEDS
The Raspberry Pi has a somewhat rudimentary USB port hardware. Our group (and many others) has experienced random failures (lock-up) of the Adafruit USB - Raspberry USB port combination with the baseline high-speed (USB-2.0) speed of the Raspberry Pi's as they come delivered.

Our experience, and many others', is this can be completely eliminated by dropping the USB port speed back to USB 1.1 speeds by adding (appending) the following statement to the file /boot/cmdline.txt

dwc_otg.speed=1

The disadvantages of this modification are that a) diskcopy speeds will greatly diminish; b) some keyboards/mice will freeze (hence the recommendations above for the Logitech device). If your keyboard/mice freeze, you can still get into your system if it is on a network and you can access it using telnet, SSH, or VNC.   As a result of this issue, I tend to keep some additional files in the /boot directory for use as needed: `cmdline.txt.fast` and `cmdline.txt.slow`.

SYSTEM VULNERABILITIES

Immediately upon creating your Raspberry, you should change the passwords for user pi and user root. The command is "passwd." The only thing that saved us in our initial learning period was that I had gotten rid of the default passwords!

You should always have a router with a firewall between your Raspberry Pi and the rest of the Internet, if your Raspberry is connected (by either WIFI or ethernet 10Base T cable) to your home network.

Both Windows and Linux machine express their TCP/IP connectivity at their IP number as a series of "ports" numbering into the 64,000s. Some of these ports are traditionally used for certain functions, and are well-known to all hackers. Here is a short list of common ports:

Table 10-2. Common Ports

Port Number	Service
13	Daytime
17	Quote of the Day
20	File Transfer Protocol
21	File Transfer Protocol
22	Secure Shell (SSH)
23	Telnet
25	SMTP
53	DNS
110	POP3 (post office protocol 3)

You should never "port forward" through your home router/firewall any service on your Raspberry on a "well-known port." If you do (as I did unaware of the risk) your Raspberry's "open port" will quickly be discovered by "bots" who will then relentlessly try brute-force attempts at guessing passwords, etc. If they gain access to your Raspberry.....

One solution to the problem, should you need to have remote access to your server (for maintenance, etc) is to have your router do a port shift -- perhaps forward a randomly chosen port number like 11045, to the SSH login port 22. This decreases the chance that your port will be discovered (but does not eliminate it).

Another option is to change the port number to which an important (and vulnerable) service, such as sshd (secure shell daemon) listens. Here is a snippet of /etc/ssh/sshd_config in which the listening port for the ssh server has been changed to Port 23123: (Note, this is NOT what mine is set to.)

```
------------------------------------------------------------------
# Package generated configuration file
# See the sshd_config(5) manpage for details

# What ports, IPs and protocols we listen for
Port 23123
# Use these options to restrict which
interfaces/protocols sshd will bind to
#ListenAddress ::
#ListenAddress 0.0.0.0
Protocol 2
. . . .
------------------------------------------
```

Log Files
Log files, including auth.log that shows who is trying to log into your system, are kept in
         /var/logs
These log files can be useful, but they can also grow to be quite large. From time to time, check the disk usage with the command df and if excessive, check on /var/logs and delete some of the log files if you don't need to examine them.

You're Finished!

That's It. Your system should now work. As you read and learn more, you can edit your LINBPQ node's configuration file (keep a backup!) and add more features and learn how to get even more out of it. It has amazing powers to connect digital streams from one radio to another, to WINLINK, from one frequency to another, out many different protocols, including PACTOR, WINMOR, PACKET, etc.

We owe a lot of thanks to John Wiseman G8BPQ for writing this for us.

Dual frequency Raspberry AX.25 Node

You may wish to run two transceivers on the Raspberry, to cover two different frequencies, for example to connect a set of community radio nodes, to a statewide backbone system operating on a different frequency. One solution to this is to employ the TNC-Pi, which can connect to the I2C port and allows "stacking" of multiple TNC's. This has been very successfully accomplished by the "Terrestrial Amateur Radio Packet Net" (TARPN)[71] and many others, but I haven't succeeded at getting the TNC-Pi to work for me (yet). I have also found no way to run more than one Direwolf-soundcard-based tnc-equivalent on a Raspberry successfully. (I have used two different Raspberries, each running a copy of Direwolf and connecting to its own soundcard, and then referenced both systems successfully using tcp/ip numbers/ports in linbpq to build a two-frequency system). **A simpler way to have a two-frequency (or possibly more!) system is to use one or more TNC-X USB-based TNC's connected to a Raspberry Pi.**[72]. One TNC-X peacefully coexists with one Direwolf-based soundcard TNC emulation on several different dual-frequency stations that our group has installed. The TNC-X is an off-the-shelf unit that is available in both kit and finished forms, and simply plugs into a usb port of the Raspberry Pi. It doesn't get much easier than that!

REMINDER: Setting up a linbpq system is no small feat. The author of this book offers to help you, as discussed at the beginning of this chapter.

11 HOMEBREW SOUNDCARD INTERFACE

Fig 11-1. *Homebrew soundcard interface mounted in an empty tea-tin. The $5 Adafruit USB audio adapter is the white object with 2 jacks, affixed to the right vertical side of the container.*

To make a packet TNC, or to send PSK31 or any of several popular digital-over-audio modes, all you need is

- a sound card that generates the audio tones,
- a volume control to match those to the transceiver mic input;
- a volume control to match the transceiver audio output to the sound card mic input,
- and a way to activate the push-to-talk circuity fast enough to follow the ARQ handshaking of modes like packet.

Add free software such as UZ7HO soundmodem.exe, or FLDIGI, and start having digital fun!

This chapter describes the construction of a simple push-to-talk vox-type circuit that combines with a USB-port sound card adapter Adafruit Audio Adapter & interface software to make a complete sound-card-based TNC using freely available software. (More work, but a lot cheaper than a TNC-X.) This circuit approximates the functions of the vox circuitry within the very popular Signalink-USB (and its competitors such as the MFJ-1204, which are growing as more people take up these modes). The circuit is not original with me; I found the basic idea on a British ham's web page. The circuit itself can be built compact enough to fit into an Altoid tin -- or the entire thing including the Adafruit Audio Adapter can fit within a tea tin, or a 4" metallic electrical junction box from a home-improvement store; or even easier, taped to the insides of a small cardboard box (e.g., granola bars box) and covered with aluminum foil.

> Note: Although technically a "TNC" is a self-contained hardware product such as the Kantronics KPC-3, in this chapter and elsewhere I'll take the liberty of expanding that definition to include a simple soundcard+electronic push-to-talk system (which requires software to complete the functions of a true TNC) as a form of "TNC" as well. Since the device described in this chapter was constructed for little more than $10 on perfboard when we first started using it, elsewhere in this book it may be referred to as the "$10TNC."

Simple, inexpensive USB-based soundcard adapters that will work with this system include the following devices -- all of which are known as "classless" soundcard adapters that comply with a standard that requires no additional drives with either Windows or Rapsberry Pi computers:

- Adafruit 1475 Audio Adapter (the one featured in this chapter), normally $5 and available directly from Adafruit as well as through other channels: https://www.adafruit.com/product/1475

- Sabrent USB External Stereo Sound Adapter (normally about $6.50) https://www.amazon.com/Sabrent-External-Adapter-

Windows-AU-MMSA/dp/B00IRVQ0F8

- Cosean USB External Audio Soundcard Adapter.
  https://www.amazon.com/gp/product/B01890C5T6/

In the first version of this book, this chapter went into great detail on building this interface circuit on standard perf-board, a process that takes an experienced builder about 2 hours. Inexperienced newer hams in my neighborhood took 3-6 times longer due to their lack of any real experience at components, construction, or soldering. A very devoted group of hams spent literally TWO SATURDAYS learning how to build simple circuitry with this project.

PRINTED CIRCUIT BOARDS

Since that time, I learned how to make circuit boards with an internet- and mail-order firm, ExpressPCB[73]. Their free software and relatively easy build process proved to be a huge win-win for me -- I was able to build the circuit board easily and have any number built and delivered. My construction time was literally cut from 2 hours, down to 30 minutes. It would be a HUGE improvement for less-experienced builders.

However, printed circuit boards aren't cheap, so the price for the entire project from from the $10 level to around $40 depending on enclosure, connections, etc.

For a limited time (until I run out of boards) this may be offered as a kit for sale on eBay. Search for the author's name, and "Soundcard Interface" or similar on eBay. Due to shipping and Ebay charges, the price had to increase but is still cheaper than a commercially available interface --- though it does require construction!

As a result of that development, this chapter will still include information on the original method of building, but will also include information on the printed circuit board, and how you can obtain the construction files, as follows:

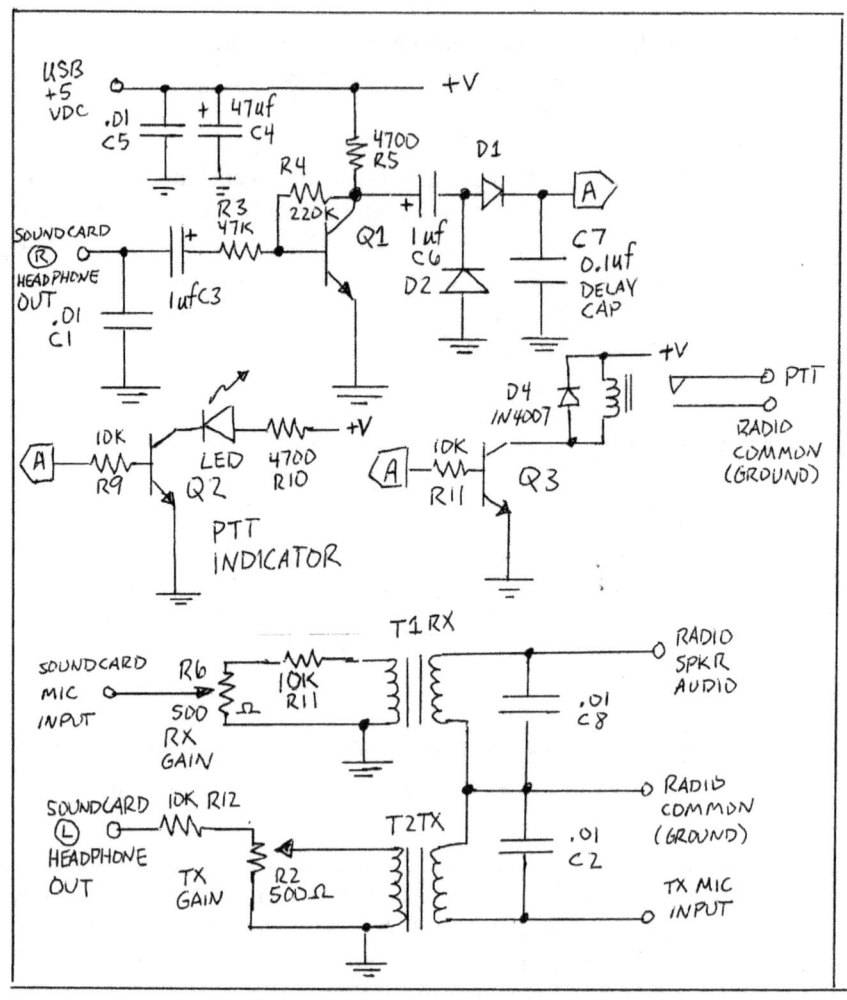

Fig 11-2. *Schematic diagram of $10TNC circuitry. Resistors are in ohms, K=1000. Capacitors are in farads, mfd = microfarads. Transistors are all 2N3904. Resistors 1/4 or 1/2 watt. Capacitors without polarity are ceramic. Capacitors with polarity marked are electrolytic.*

For the amateur radio club or group that wishes to turn this into a project, I'll gladly send them (by email) the .pcb file that allows them to order their own boards. While finished boards start at about $20 in small quantities, they drop to $10 with large quantities. (Contact me if you would like to join in an order.) The parts for this kit can be obtained under $10. A simple steel electrician's house wiring 2-gang junction box can be had for just over a buck, and a cover for just over 50 cents. Makes for a very inexpensive shielded box! With an experienced and patient mentor, this would be a great project for helping amateur radio operators with little previous construction experience, learn what fun it can be to actually *build your own equipment.*

Contact me at: docvacuumtubes@gmail.com

Fig 11-3. *The Printed Circuit Board.*

Materials

ITEM	DETAILS
Adafruit Audio Adapter (USB-based sound card)	$5 https://www.adafruit.com/product/1475 also try https://chicagodist.com/products/usb-audio-adapter-works-with-raspberry-pi
Prototyping Printed Circuit Board	Small board -- http://www.mcmelectronics.com/product/DISTRIBUTED-BY-MCM-21-113-/21-4590
500 ohm potentiometer, R2, R6	https://www.digikey.com/product-detail/en/bourns-inc/3306K-1-501/3306K-501-ND/84791
4.7K R5	https://www.digikey.com/product-detail/en/yageo/FMP200JR-52-4K7/4.7KZCT-ND/2059014
10K, R1, R7, R9, R11	https://www.digikey.com/product-detail/en/stackpole-electronics-inc/CFM12JT10K0/S10KHCT-ND/2617547
47K, R3	https://www.digikey.com/product-detail/en/stackpole-electronics-inc/CF14JT47K0/CF14JT47K0CT-ND/1830391
220K, R8	https://www.digikey.com/product-detail/en/stackpole-electronics-inc/CF12JT220K/CF12JT220KCT-ND/1830532
Ceramic 0.01 microfarads (10,000 pf) C1,C2, C5, C8, C9. Ceramic, any voltage 25 or greater.	https://www.digikey.com/product-detail/en/murata-electronics-north-america/RCER71H103K0K1H03B/490-11884-ND/4277785
Ceramic 0.1 microfarad C7.	https://www.digikey.com/product-detail/en/vishay-bc-components/K104Z15Y5VF5TL2/BC1160CT-ND/286782
Electrolytic (polarity important) 1 microfarad C3, C6, any voltage >15 volts	http://www.digikey.com/scripts/DkSearch/dksus.dll?Detail&itemSeq=228518274&uq=636324138638111468

47 microfarads, C4, any voltage >15 volts	https://www.digikey.com/product-detail/en/panasonic-electronic-components/ECA-1EHG470/P5539-ND/245138
2N3904 transistor, Q1, Q2, Q3	https://www.digikey.com/product-detail/en/fairchild-semiconductor/2N3904BU/2N3904FS-ND/1413 2N3904 pinout: http://www.futurlec.com/Transistors/2N3904.shtml
1N4148 or 1N4448 (or virtually any diode)	Use any small signal diode or 1N4007's if you have none.
1N4007	https://www.digikey.com/product-detail/en/fairchild-semiconductor/1N4007/1N4007FSCT-ND/965481
RL1 5volt, reed relay, low current draw (<20 mA)	https://www.digikey.com/product-detail/en/coto-technology/9007-05-00/306-1062-ND/301696 (10mA coil current)
LED	https://www.digikey.com/product-detail/en/tt-electronics-optek-technology/OVLBR4C7/365-1175-ND/827111
Ferrite Core for transceiver cable	https://www.digikey.com/product-detail/en/laird-signal-integrity-products/28A2025-0A0/240-2074-ND/242802
1:1 600 ohm audio transformer	http://www.digikey.com/product-detail/en/tamura/TTC-5028/MT7212-ND/674291 (close enough to 1:1 to work) or http://www.ebay.com/itm/2pcs-600-Ohm-Audio-Telephone-Radio-Coupling-Isolation-1-1-Transformer-US-Seller-/321917511674 or http://www.mcmelectronics.com/product/287-1404 Beware of shipping charges. Much cheaper to

	purchase a larger lot (on ebay) and wait for international shipping if you have a larger group of builders in your group!
LED	http://www.digikey.com/product-detail/en/cree-inc/C503B-RCN-CW0Z0AA1/C503B-RCN-CW0Z0AA1-ND/1922930

Tools

- small needle nosed pliers
- small diagonal cutter ("dikes")
- low-wattage soldering iron (e.g., 25-40 watts) Here's an example from Walmart: https://www.walmart.com/ip/Nippon-America-71B30KT-30W-Soldering-Iron-Pen-Kit/49016054 for under $10.   It is almost impossible to build this circuit with a big soldering gun -- a nice pencil point gets into tight spaces and doesn't create "solder bridges"
- something safe to set the hot iron on, that will cradle it to some extent. metal, ceramic or brick. I use good sized electric junction box.
- rosin core (electrical) solder, lead alloy preferred (lower melting temperature). You won't need a lot.   I prefer the 60/40 lead tin (lowest melting point) but it is getting harder to find. It is a lot easier to build small circuits if you have slender (small diameter) solder
- Optional: alligator clip parts holding "third hand"
- Optional: magnifying glass
- Voltmeter ($5 Harbor freight digital multimeter works fine)
- You may want to have a "solder sucker" or "solder wick" to absorb misplaced solder that creates an unwanted solder bridge.

CONSTRUCTION

NOTE-- for the printed circuit board version there is a complete construction booklet available on Amazon: https://www.amazon.com/KX4Z-Sound-Interface-Construction-Manual/dp/1545480079/--- and a free version (PDF) at this URL: http://www.qsl.net/nf4rc/SmallBookSoundCardConstructionManual2.pdf and for a limited time there will be a full kitted product available on Ebay -- http://www.ebay.com/itm/-/263023729596? , or search for "Ham SoundCard Isolated Interface" to find the kit.

<u>For standard perf-board builders:</u>
Because this is audio frequency construction, lead dress isn't all that important.

This circuit has been built by several people in Gainesville Florida, mostly retired volunteers who had little previous building experience. Even with very significant coaching, these builders took many hours to complete, test, and hook up this project. By comparison, an experienced builder in our area completed a beautiful circuit in only 2 hours, and my times are similar.

Circuit Description

The audio signal from the USB audio device is a low voltage audio signal, less than 1 volt. Q1 is a simple transistor amplifier that develops roughly a 4-volt peak-to-peak AC signal from the low voltage audio input signal. Diodes D1 and D2 rectify this to create a positive DC voltage. C7 provides a very small delay to keep the transmitter ON. The turn-off delay constant of this circuit is at a minimum as a result of 0.1 microfarads and 3.3kohms (if all three output circuits are built), roughly 0.3 milliseconds, but probably 2 time constants have to expire before the PTT stops. This is long enough to keep the push to talk energized throughout the audio signals. The DC voltage developed across C7 is then used to drive a transistor that controls a relay to close the push-to-talk circuit of the transceiver. Two 600 ohm 1:1 audio transformers are used to completely DC-decouple the radio transceiver from the remainder of the circuitry RF filter capacitors (0.01 microfarads) are liberally sprinkled in strategic locations to avoid RFI. Note that the radio common (ground) is NOT connected to the computer ground.

In this text, I can't hope to give enough guidance to fully meet the needs of a person with zero previous circuit construction experience; so I'll just provide a modest discussion to give an idea of parts layout.

The board suggested for this project has two "bus bars" running across the center of the board, however these are really CLOSE to each other to use one as the plus voltage and one as the negative....too close for possible soldering errors. So I use one of them as the positive bus bar and run a bit of hook up wire along the periphery of the board (both top and bottom) as the "ground". Although I originally put it on the component side, I think it would actually be better to put the ground bus on the printed side.   Either can be done.

In this case, I built the input circuit on one half of the board, separated by the central +V bus bar, and then rotated the board 180 degrees and built the desired output circuits on the other half of the board.

I recommend one start first with the input audio amplifier Q1 and the two-diode rectifier circuit that develops the DC voltage used to drive the PushToTalk relay.   Using a 9-volt battery (watch polarity!) to power the circuit, check the collector voltage of Q1; it should roughly be between 2 and 3 volts. You do not want to use your computer as the power supply for this test, due to the risk that you might have an unexpected short circuit.   Q1 should not be saturated (<0.6 volts) nor driven into cutoff (rising to supply voltage).   If either of those is true, something is wrong with the DC biasing or connections around Q1 and must be corrected before proceeding further.

Wiring to the USB audio dongle:

There are four signals you need to connect to.. There are two alternative methods; either works well.

Method 1. Crack open the case of the USB audio adapter and solder to the individual connections.
Method 2. Use a separate USB cord (cut in half) to get the +5 VDC, and use two stereo 3.5mm plugs/cables to gain access to the sound card audio signals.

If you choose method 1:

Using tiny very flexible stranded wire will make this a lot easier. After you get all the connections done and have tested the circuity, I recommend a bit of 5 minute epoxy to glue the USB circuit board back into the bottom half of its clamshell, and a dab on the short white cable coming from the computer to strain relief it.

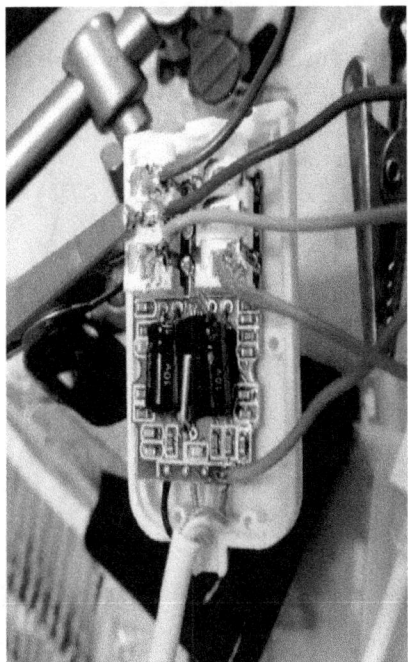

Fig 11-4. *Wiring directly to the USB Adafruit sound dongle. Not for the novice solderer. The connector with the three wires is the headphone output (TX signal); the connector with only one wire is the mic input (RX signal). The wire at the bottom edge of the board picks off +5.*

- +V (5V DC) which is on the little red wire coming from the USB. You are probably going to have to disconnect that tiny wire, connect to it, and then run a short jumper back to the board. This is tiny, tiny soldering so be careful.
- Ground -- easiest picked off of the farthest tab on the audio connectors. Can be found on either.
- Audio output from the computer --- from the channel closest to the computer
- Mic input to the Computer (will go to the speaker output from the transceiver) -- from the tab closest to the computer.

On both the audio output (TX signal) (left in the photo) and mic input (RX signal) (right), the terminal closest to the edge of the device is ground, and the most- inward terminal is the signal you want. Solder quickly so you don't melt the entire jack and it will work nicely!

If you choose method 2 (use stereo 3.5 mm cables):

Headphone plug (TX Signal):
TIP =   Left Channel
RING =   Right Channel   (used for PTT)
SLEEVE =   GROUND, both DC and AC

Mic Plug (RX Signal):
TIP =   mic input
RING   (not used by us)
SLEEVE =   GROUND, both DC and AC

Wiring to the Transceiver:

There are four signals you will need to connect to the transceiver:

- Mic input
- Speaker output
- Push To Talk control
- Ground

I recommend using shielded wire to run these signals to the transceiver, with the ground connected to the shield at the computer/circuitry side.   Leave enough room to put a couple 2.5" loops in this cable to add common-mode inductance, and also clip on a ferrite bead close to the transceiver.

In my station, all cables coming from transceivers terminate in a RJ45 wired in the way that would connect to a Signalink which has been configured for their Baofeng/Kenwood connector (as previously discussed in this book):

RJ45 Pin	Signal
1	Mic
2	Ground
3	Push to Talk
5	Speaker audio

I chose to wire the cables from my TNCs (the ones that don't have female jacks accepting RJ45 like a Signalink) in the exact same way, and then use a female-female RJ45 compatible adapter to connect TNC to transceiver. This makes it a lot easier for me to "mix and match" different radios and TNCs when necessary.

Fig. 11-5. *Built and installed Version 1.1 (current version) printed circuit-board based sound-card interface. The gray cable going off to the top is shielded 10baseT twisted pair cable going to the transceiver.*

SOFTWARE for WINDOWS USAGE

I use UZ7HO soundmodem.exe (the 1200 version, not the high speed 9600 baud version) for packet. This wraps the sound card/ptt circuity and gives it both an AGWPE and KISS tcp interface. Version 0.97 is current as of this writing. http://uz7.ho.ua/packetradio.htm

If you add UZ7HO's excellent EasyTerm software, which uses the AGWPE interface of the soundmodem.exe driver, you'll have a full "classic" packet interface that will allow you to connect to nodes, participate in chat rooms, and other standard packet activities -- also available on his web page at: http://uz7.ho.ua/packetradio.htm

A commercial alternative to the UZ7HO software, is MixW, which provides for not just AX.25 packet but also most of the popular HF modes http://mixw.net/ (there may be a charge for purchase). It does not require soundmodem.exe; like FLDIGI products, it includes all of the required connections.

You can also use (free) FLDIGI with this board for modes such as PSK31 and RTTY.   http://www.w1hkj.com/

If the ptt delay is too short, increase the value of C7. I think this might only be a problem if you were using this for CW, and needed to keep the PTT engaged for proper operation.

This circuit works well with WINLINK, using UZ7HO soundmodem.exe to provide a KISS interface. http://www.winlink.org/ClientSoftware (see WINLINK EXPRESS download link, lower down the page)

Setting the Volume Controls

If you're using a laptop, be sure to set the speaker volume output to this device at maximum. If you're using a Raspberry PI, see the section below to learn how to use `alsamixer` **to set the volume levels of input and output.**

1.   Set the RX volume control in the middle of the range where your software properly decodes some normal-sounding beacon or packet station.
2.   St the TX volume control (with your computer or Raspberry PI set to 100% output volume) by monitoring your transmitted signal, and setting the TX gain just below the point where the received audio no longer gets any louder (on HF, still in the linear range; on FM, within the normal deviation range).

Setting RASPBERRY PI Audio Gain (Volume) Levels

You'll need to set the mic gain and audio output levels if you're using a soundcard-based technology. If the audio output isn't at 100%, Signalink and similar devices (such as the $10 TNC) may not key the transmitter properly. From the command line,

```
alsamixer
```

will bring up a pseudo-graphical volume control. Use Left/Right cursor keys to reach your sound-card device. Use the up/down arrow keys to adjust volume. You probably want the audio output at maximum and the mic gain at 50%.

Note that the `alsamixer` works perfectly with the Adafruit USB audio adapter—and with other "classless" soundcards.. (Unfortunately as of this writing, the popular Tigertronics Signalink doesn't seem to allow setting of all the gain controls with `alsamixer`, making satisfactory permanent adjustment of the Raspberry volume levels more problematic; this may be due to the manual potentiometer adjustment of those levels, however I'm not yet able to be certain that settings will remain stable. The MFJ-1204 competitor does appear to interface correctly with `alsamixer`.)

After you get your signal levels the way you want them, you'll need to use this command to lock them into place so they remain after the next reboot:

```
sudo alsactl store
```

Setting Windows Laptop Audio Gain (Volume) Levels

If you're using a Windows computer, (e.g., with FLDIGI, MixW, BPQ, Outpost or other software) you can adjust the volume controls as with any other windows component.

Figure 11-6. *Printed circuit board mounted (actually wedged) into a 4"* *electrical box. One knockout has been removed to allow access to the* *potentiometers, The USB dongle is also inserted, and another knockout* *has been removed and a "3/8 non-metallic clamp" used to secure the* *wires' exit point. There is a standard solid metal plate that fits on top as* *a cover.*

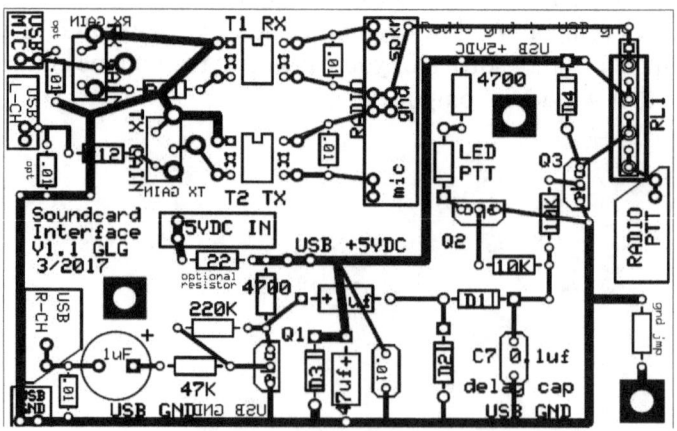

Fig. 11-7 The printed circuit board. Reversed writing are copper *lettering on the opposite side.*

12 STORE AND FORWARD SIMPLEX VOICE REPEATER

If duplex repeaters are non-functional due to loss of power, tower/antenna damage, EMP/lightning, or any other cause, having a spare store-and-forward simplex voice repeater can be a huge help. Placing it as high as possible will give your community a simple means of VHF/UHF repeater function, albeit at a slower throughput than a traditional duplex repeater.

This is because a simplex store-and-forward repeater works by capturing a voice transmission, and then simply replaying it back over the same frequency.

A commercially available digital device that plugs into a transceiver's headphone output (to capture the incoming signal) and microphone input (to play it back) is the Surecom Sr-112 Simplex Repeater Controller. With a cable that fits Kenwood/Baofeng transceivers, it is available at: https://www.amazon.com/Surecom-Simplex-Repeater-Controller-Kenwood/dp/B014H2D2TM This little device works well, includes its own li-poly battery, and can be powered with 5-24 VDC. It could easily be powered from a storage battery along with a 2-meter transceiver. The cable handles the usual 3 signals -- mic, receive audio, push-to-talk -- and I was easlily able to connect it to one of our standardized RJ45 pinout plug systems, so now it can be connected to 2-meter rig in our group.

NOTE: the multiconductor audio plug must be FULLY inserted into the jack on the back of this device....if things aren't working properly, check that first! (Ruined a demo, once!)

This is a great little unit to have in your equipment cache for emergency replacement of an important repeater.

Fig 12-1. *SURECOM SR-112 Digital Simplex Repeater connected to Baofeng transceiver.*

Instructions for using a Raspberry Pi to function as a simplex digital repeater (with a suitable soundcard and PTT circuit) are given at: http://www.twotonedetect.net/raspberry-pi-simplex-repeater-the-peaterpipyr/ ; however these were based on an earlier version of the Raspberry Pi and I have been unable to make them work with a Raspberry Pi Version 3.

13 WINLINK SERVER STATIONS

WINLINK, which is a radio-based email/attachment system, is a mauture, two-decades-old project that has really become a defacto standard for backup / emergency / high-seas email solution. In the month before submission of this manuscript, on HF alone the WINLINK system handled over 25,000 messages, comprising over 111 Mbytes of email. That is an astounding amount of traffic to be handled over HF amateur radio. Further, it's an all-volunteer group! Technically it is managed by the Amateur Radio Safety Foundation, Inc., a worthy charitable organization.

A small cadre of Development Team members keep churning out improvements and keep their systems working with an ever-expanding list of ham radio gear.   It's a pretty amazing group. As part of their support network for users, they provide and manage a few email groups which allow for interchange of information beyond that on their web server (www.winlink.org). Those groups are currently hosted on Google Groups.

WINLINK has a significant array of software products/solutions. Additionally, John Wiseman G8BPQ has both contributed to the WINLINK system, and has complemented it with his BPQ line of ham-free-usage software featured highly in this text.

The purposes that WINLINK serves include:

- Free email access for pleasure or emergency on the high seas, utilized by a significant number of sea-going vessels, and with a notable history of SOS messages handled.
- Free email access to anyone else who want/needs it -- this includes people in rugged areas, as well as people anywhere in the world.
- Email backup for U.S. government systems, as a backup for

cyber-warfare damage to normal communications systems, through the SHARES program.
- Free Email for emergency communications, used by many ARES, Red Cross and EOC groups.

WINLINK works by having "gateways" that can accept and disburse email over either HF or VHF frequencies using automated computer control, and then either forward that email over the Internet or over HF radio (should the Internet be unavailable). These gateways are known as Radio Message Servers (RMS). Multiple online presentations and YouTube videos are readily available to help to better explain a rather comprehensive and complicated system.

VHF Packet gateways typically are fixed on one frequency and gateway from radio to Internet (but may also connect to an HF server, see below).

HF RMS Server gateways typically are set to sequentially monitor several frequencies, usually on different amateur bands, continuously & sequentially changing monitor frequencies every few seconds, 24 hours a day. They additionally have the ability to command the transceiver to move to a totally different frequency to make a connection to a 2nd server so as to transfer email by radio, on a frequency known to be monitored by the other server. This is quite a complicated system!

At the time of this writing, HF RMS servers are able to use multiple protocols for email transactions -- including

- WINMOR, a protocol allowing the use of inexpensive soundcard technology, but unable to forward email; and
- PACTOR, a proprietary protocol on fairly expensive high performance radio modems designed for amazing speed on less-than-perfect HF band conditions.
- ARDOP – a new protocol in development.

For the purposes of emergency communications, one wishes to be able to maintain communications both across a city or county, and also to be able to reach outward, to state assets, or federal assets in time of real emergency. Such was the situation during Hurricane Katrina!

Packet-based VHF RMS stations are common across America and

provide easy connection to the internet from ham radio station using free WINLINK EXPRESS software and any kind of Packet connection.

However, the biggest need for communications is precisely when the Internet is *out of service in an affected area*! That is where the HF RMS stations become key.

A "Hybrid" RMS server is able not only to connect to the Internet, but is able also through RMS_RELAY software, to cache and forward email over HF radio frequencies, provided that the station is computer controllable and equipped with a PACTOR modem.

With one "hop" the email may be sent by a RMS server in a disaster area directly to a second RMS server that still has Internet access -- and the email then moves almost immediately to the intended recipient. Candidly, the process of radio-forwarding however is slow and fraught with difficulties --- it is attempted only at set time intervals, and any interference (QRM) on the known frequency of the intended recipient station causes the attempt to be abandoned. Email may forward in 15 minutes on the very first attempt....or it may take more than twelve hours in a really difficult situation to get out....

Adding human oversight of this system can dramatically speed up the process; in my experience, I can quickly recognize when an automated contact attempt is doomed to failure, and shove the system along toward a better connection.

It's important to point out that a community with multiple experienced HF hams, using WINLINK EXPRESS (client) software, can move a lot of email in and out of a community, by connecting directly to distant RMS servers – the strategy wisely chosen by the Marion County Hospital Emergency Communications group. Meanwhile, a local VHF-based WINLINK gateway can be utilized for intra-community connections (the alternative is packet file transfer using YAPP, or tcp/ip based MESH connections). However, if your group doesn't yet have enough HF experienced hams (our current situtation) --- you'll be more dependent on VHF-to-HF gateways, which work, but slower than an experienced human.

VHF PACKET RADIO SERVER
|
|
RMS_PACKET software on Windows
|
|
HF HYBRID WINLINK RADIO SERVER

Figure 13-1. *Connection between VHF Packet server and HF Hybrid Server.*

The real power of the system becomes evident when the RMS station includes not only Packet access but also HF access, and with the two connected with RMS_RELAY. Every possible connection is then is possible:

Local email from emergency officials within the city/county can be received on VHF packet, cached, and disbursed back to the proper person just like any normal email service, by RMS_RELAY.

Email from local emergency providers addressed to distant recipients can be accepted on VHF packet, sent out over HF radio; replies received and disbursed over VHF packet.

Obviously, it is a real advantage to a community to possess one or more "full service" RMS stations that provide both VHF and HF services. And the best of all worlds is to also have a cadre of experienced HF hams operating WINLINK client software, handling long-distance email with distant RMS stations in unaffected areas.

WINLINK HYBRID SERVER STATION HARDWARE
To achieve this HF forwarding, the server station must include the following equipment:

1. Computer controllable transceiver
2. PACTOR modem
3. Optionally: soundcard interface allowing WINMOR protocol

4. Combiner box to allow both interfaces to function
5. Multi-band antenna, potentially utilizing an automated antenna tuner, or alternatively a broadband or multiple-dipole array.

WINLINK RMS SYSOP COMMITTMENTS

In order to be given access to the WINLINK CMS (central message server) servers, a prospective systems operator ("sysop") commits to making a sincere effort to have their station remain available 24/7/365, and additional to monitor required email group(s) to keep abreast of system upgrades and changes.

For more information on becoming a WINLINK systems operator see:
http://www.winlink.org/content/join_gateway_sysop_team_sysop_guidelines

CLIENT LEVEL WINLINK

There are certainly many more users (clients) of the WINLINK system than there are servers! Client access is the first key to maintaining email access even when normal telecommunications are non-functional. VHF packet client access will be primarily useful only if you can reach a "full service" or hybrid RMS station. However, HF client level access gives you access to HF RMS server stations anywhere in the world! This should be a highly prized goal for all emergency communications groups.

Furthermore, there is specialized WINLINK software that allows client level access to amplify its usefulness by creating a local (pop3) email server, and piggybacking non-ham user accounts on top of access by a licensed amateur radio operator. The WINLINK software package that does this function is known as PACLINK. There is a Unix 3rd party software that runs on a Raspberry Pi with similar function. [74]

WINLINK & CONFIDENTIALITY

Amateur radio operators are not allowed to encrypt transmissions. Thus there is no complete confidentiality on WINLINK emails. Further, sysops are able to review email transitioning their servers (in order to meet FCC requirements for prudent management). However, in practice, there is much less chance an email transitioning the WINLINK system is going to leak important information to the wrong ears. First, the message transmission is not "open to the world" as are "broadcast" protocols. The WINLINK system uses high speed error-corrected handshaking systems on the radio which are difficult to intercept even by people who understand them. Further, the systems operators are generally high quality volunteers who are far too busy to go snooping through the email that moves through their system. My system typically moves over 500 kilobytes of information every three weeks; I'm not spending my time reading all that email!

In a recent online forum discussion, the HIPAA confidentiality issues were discussed and one knowledgeable ham pointed out two important points: security of transmission systems for patient records is an "addressable" standard --- meaning rather than encryption being an absolute standard, organizations had to "address" (give reasons for) their security practices.[75]

In an emergency, groups might well be able to defend transmitting even individually identifiable patient data by a less-than-totally secure system such as WINLINK....when there aren't any alternatives left! Furthermore, many other sections of HIPAA requirements seem to have "loopholes" for emergencies, which could be pointed to if there was a question.

No one will know for certain, of course, unless there is an authoritative legal opinion or a court case involving "PHI" (protected health information) transmitted over a non-encrypted system during an emergency. Amateurs should advise agencies of organizations they serve of the characteristics of their systems, and document the data and messages they are asked to transmit/receive in appropriate recordkeeping. **And of course, don't KEEP any patient information on any computer after it is no longer needed there!**

14   WIRELESS VHF NETWORK DEVELOPMENT

First, the bad news for VHF Users.   While 31-Hz narrowband PSK31 signals on the HF bands enjoy a significant advantage watt-for-watt over almost all other signals, 1 kHz wide packet signals on VHF are still stuck with the full FM bandwidth but fail if just a few bits are misread, and hence perform poorly for distance-awards when compared to FM voice. The power-per-hertz-of-bandwidth greatly favors PSK31 as compared to AX.25 Packet. Further, voice can still be somewhat recognizable & useful only a tiny bit above the noise, whereas missed bits in a non-forward-error-corrected packet signal means the entire packet is rejected and must be re-transmitted.   You are unlikely to get near the range of simplex voice when you are planning for VHF packet, unfortunately. And while broadcast modes like MT63 can successfully send bulletins over a voice analog repeater in a pinch, rapid back-and-forth ARQ Packet burps simply can't utilize a voice repeater at all.

Estimating the range of a station limited to line-of-sight in any environment other than a mountain peak transmitting through free space is fraught with error. The best that can be hoped for are the limits imposed by **free space loss** and the **curvature of the earth**.

Free Space Loss Estimation
Free space loss in decibels can be approximated by

$$FSL(dB) = 20 \log_{10}(d) + 20\log_{10}(f) + 32.45$$

where   d = kilometers   (Note: 1.6 km/mile)
             f = Megahertz

Note doubling the frequency or the distance adds 6dB to the free space loss.   The distance effect is due to the spreading out of the wave, so that the power density continuously declines.   The frequency effect is *not* because higher frequencies are less effective; it is because the formula is based on an isotropic antenna, and as the frequency increases and the size of the antenna decreases commensurately, the aperture declines and thus the energy extracted from the radio field declines. Of course, it is also much easier to make higher gain antennas (effectively, returning the aperture toward that of a lower frequency antenna) for higher frequencies.

One may start with the transmitter power in watts, convert to dBm (dB relative to milliwatts), subtract the free space loss, and compare to the expected sensitivity of their receiver.   Modern FM voice receivers will have a sensitivity in the range of 0.5 microvolts or -113 dBm.   A transmitter producing 25 watts produces 44dBm.   If we assume 4 db of transmission line loss and isotropic antennas on each end, then a link budget

Transmitter Output:	44 dBm
Transmission line loss	-4 dB
TX Antenna loss:	0 dB
RX Antenna loss:	0 dB
RX transmission line	-4 dB
Receiver sensitivity	-113dBm

would indicate the maximum path loss allowable is 149 dB.

Unfortunately, <u>seldom do radio waves enjoy the low losses calculated by the Free Space Loss.</u> Losses in urban / suburban / building environments may exceed 10-20 db PER MILE. That eats up a 149dB path budget rather quickly. And furthermore, for digital packets to arrive unscathed, the S/N ratio often has to be rather large. In my experience it is common for VHF signals of 20Watt transmitters at vehicle roof level to peter out within 3-4 miles.

Curvature of the Earth

Radio waves just don't go through dirt. So the curvature of the earth imposes another very significant limitation. The approximate distance from each antenna until a horizontal wave from it hits the dirt is called the **radio horizon** and is variously estimated around 1.4 x square root of height of antenna in feet. If you wish to be optimistic, use the top of the antenna! A table in the Reference chapter gives maximum distances possible based on no-hills earth. In the case of hills/trees etc., diffraction allows some passage of signals that would otherwise not be expected, but with significant losses.

Our Alachua County ARES group has found reasonable estimates of range to be given by the free online Radio Mobile mapping tower coverage simulator at http://www.cplus.org/rmw/english1.html For digital coverage maps we typically adjust the requirements to 90% accurate signal reception and boost the requirements for "strong signal" to 12 dB increase above that. Even then, the results are only approximate.

High quality topographic durable printed maps can be obtained (at a cost) from http://www.mytopo.com/index.cfm In my experience, a 1:50,000 scale is useful for examining the terrain of a region of approximately 20 x 30 miles. At this scale, the writing on the contour lines is quite small, requiring a magnifying glass. Also remember, both holes and peaks have closed contour lines! In my experience, it was most helpful to find generally the highest contour lines in my area and to highlight those, as the "tops" of the hills. This allowed a rough grasp of the BEST locations for digital repeaters, and also the general lay of the most challenging obstructions. In our rolling hilly area where elevations can vary by 100 feet within a mile, resulting in multiple diffraction events for tree-based antennas, we generally hope for 6 miles radial range from one residential tree-based antenna to the next, and we are happy when we get more than that. Of course, the range to hundreds-of-feet-high tower-mounted antennas will be significantly greater, as their signals escape the high losses of the foliage for many miles.

Working out range data for a prospective digital network is not all theory; for the final answer one must actually get out there and MEASURE signals. Here's one setup crudely constructed ad-hoc with duct tape in the back of a pickup truck to test signals (I've also hoisted antennas up trees in parks):

Figure 14-1. *15 foot tall antenna in back of pickup truck.*

When attempting to put together a city- or county-wide digital relay system, the following suggestions may be of use to you:

- Do not depend solely on just one digital repeater. Have alternates.[xiv]
- While high-tower or building-mounted antennas will likely give you much better range, they can be very difficult to repair in bad weather, storms, power outages and when all available authorities with the access (and the keys!) are busy.
- While residential tree-mounted antennas can be much more easily reached (by paying out rope to lower the antenna) by their

---

xiv I used precisely this argument when working to develop a digital residential network in my county, and within months the only previous digital repeater was indeed experiencing significant problems....and could not be reached for weeks.

owners, those owners may have a variable degree of understanding early in their ham radio careers of how to diagnose problems and how to fix them. Their attentiveness to their station may also be in competition with spouse- and family-needs.

- Complicated Linux-based repeater systems are highly reliable, which is good, because finding people who can really understand them and are willing/able to manage them in a pinch by themselves is like hunting for the proverbial hen's tooth. Consider remote access (with appropriate security) in some cases.

- It is a good idea to always provide users with some visual means to verify their transmitter is producing an output, and roughly how much. Even a cheap used HF SWR meter works.

- A 230-volt gas-discharge surge arrester costs only about $2 and may go a long way to protecting your precious digital repeater station from ESD or EMP damage.

- Linux machines are best provided with instantaneous battery backup, because they don't like losing power before the write cache has been written out to disk.

- Until people truly grasp the benefits of a widely resilient digital repeater system, you will have a lot more success if you foot the bill for most of the equipment. That will also encourage you to get the costs down!

- It is a very wise idea to always make digital equipment capable of being used for voice transmissions. At the least, this gives an inexperienced person a way to test the radio portion of the system easily.

- If you don't build your digital repeaters on some sort of supporting structure, even a plywood board, is is very unlikely that they will ever be able to be moved even across the room, without something disconnecting. And inexperienced users will find all kinds of way to mis-connect it.

Once you have a quality group with significant paperwork to prove it, don't be afraid to approach authorities or owners with significantly high assets that could provide sites for much better digital repeater installations. When our town's sole high-perch digital repeater experienced a tower antenna mount failure, people with significant expertise came alongside to assist us in not only replacing the broken mount with a homebrew unistrut-based inexpensive mount, but also we were able to install a complete new digital repeater system at the same time, with an additional transceiver, coax feedline and antenna.

ASK AND YOU SHALL RECEIVE

When I approached the Division of Forestry for temporary help with providing better signals to two hurricane shelters on the west edge of our county, we were handed indefinite usage of a pre-existing high gain vertical antenna already mounted on the top of a fire tower and outfitted with 7/8" heliax going to a protected empty building that already had electrical power --- and our digital coverage significantly enlarged.

Fig. 14-2 *Forest Grove fire tower -- and the antenna we were given indefinite usage for a digital repeater covering the western edge of our*

county.

INCREDIBLE POTENTIAL EMERGENCY COVERAGE

We actually didn't realize just what a wonderful gift the Division of Forrestry had made to us. Since that time, I've done some simulations of what it would take to provide emergency digital ham radio connections clear from the west coast of Florida (e.g., Steinhatchee) back to Gainesville Florida. (Steinhatchee and Cedar Key get hit by hurricanes with some regularity and there have been some that did very significant damage.)

A recent experiment from a similar lookout tower in the Steinhatchee area indicated that it is actually quite simple to connect back to the digital-rich center I-75 corridor of the state from such high towers in the flatter coastal areas of the state. Our local Alachua County ARES group may test such emergency communications deployments as part of an upcoming ARRL Simulated Emergency Test.

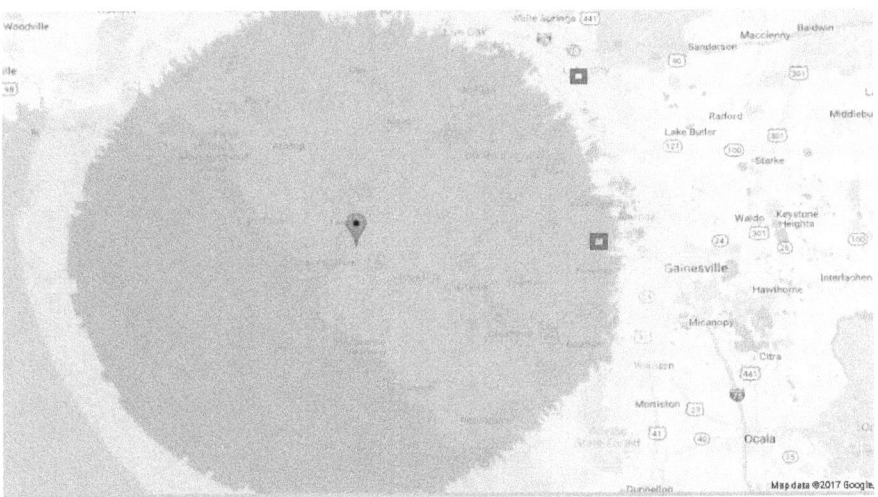

Fig. 14-3. Radio Mobile predicted good (yellow) and excellent (green) coverage from the Jonesboro lookout tower near Steinhatchee Florida. Coverage extends to both our recently added Forest Grove digital node (lower red square) and to a SEDAN-network digital node in Lake City, Florida (upper red square) and covers a very large segment of the Nature Coast of Florida. This tower has survived for well over 70 years.

15   HISTORICAL DEVELOPMENT OF MARION COUNTY HOSPITAL EMERGENCY COMMUNICATIONS

Based on a talk given by Dave Welker W2SRP, on Feb.8, 2017 to the Alachua County ARES group.

Dave Welker, W2SRP, is a medical illustrator – and previous head of such a department -- who retired and moved to Florida in 2004, right in time for some impressive hurricanes making Florida landfall.

He was one of a group of people who were approached by Marion County hospitals asking for solutions that would allow the hospitals to be able to contact state emergency officials in Tallahassee and others, if such a hurricane were to hit again.

After some study, their group settled on using WINLINK both peer-to-peer and client-server models, to provide effective VHF and HF communications. Dave came up with a proposal for the hospitals, and with hospital support, gave a presentation to the city Mayor and some Council members.   Their communications proposal was accepted, and a $12,000 grant got them started with equipment, including PACTOR modems necessary for high speed WINLINK communications.

They subsequently built an impressive group out of volunteers, serving and providing communications to five local health care facilities, each of which maintains its own incident command center. In 2017, Dave's group counts a total of 20 volunteers.

VHF figures highly in their systems, with antennas mounted as high as possible, and voice as well as WINLINK capabilities.   Some of the facilities have additional HF WINLINK capabilities, with back-to-back

"HamStick" loaded whips providing horizontal antennas on the top of roofs. (An earlier multiband vertical was hit by lightning). By connecting different pairs of single-band HamSticks in parallel, they achieve multi-band capability. At one facility they used 350 feet of RG8 coax to reach their antenna.

The most impressive part of Dave's organization (beside their impressive annual photo with a slew of volunteers) is his detailed training and assessment mechanism.

All his volunteers are fully trained and badged volunteers at the facilities at which they serve. Dave runs frequent training exercises with very impressive and detailed planning write-ups and after action reports, and he takes careful note of who participates and demonstrates which abilities at tasks. He seems to feel a responsibility to place trained and capable volunteers at locations, where they serve 12 hour shifts. Every volunteer is interviewed, has a background check, completes IS-100, 200, 700 and 800. Volunteers renew their commitments annually. In Welker's view, people who haven't attended meetings or training are a "liability." When a hospital administrator can ax your program if they see sloppy performance, it encourages accountability.

16 WINLINK AND SEDAN WORK TOGETHER

by Barry Isbelle N2DB

Winlink and Sedan are both packet radio programs but serve different purposes. Sedan in an interactive terminal program ideal for "Command and Control". Win Link works great for "Health and Welfare " messages.

In the Orlando area we use both at shelters.

The Sedan network is used by the county to direct and control the flow of information into and out of the shelter. The EOC and the remote shelters enter the Sedan Node Chat room together. The network will support 8 to 12 sites. Local Red Cross support Hams can enter also. By being at this round table members can type "SMS (short message service) type messages up to 80 letters long into their terminals. By hitting enter those messages will be displayed onto each of the other members terminals.

To illustrate how this saves time, the EOC will type in weather report. The weather report can be a whole page at once with proper carriage returns for each line. This in turn will be sent in clear text to each other shelter. This saves the EOC phone calling each shelter and having each shelter manager write the weather report long hand. The weather might change by the time they got thru phoning each shelter. Instead each shelter can just screen print the weather report out.

Another special aspect is shelter to shelter bulletins. Let's say Shelter A has to many or not enough meals on hand. By simply typing in a quick SMS line each of the other shelters are alerted to the situation. This works well with hourly shelter reports also. Each shelter can prepare their hourly report and upload it to the group. The shelters will need to coordinate the schedule to insure that they do not talk over each other. Of course Emergency Bulletins are immediately available. to each member.

Winlink serves a different need. People in shelters need to let their families know their "Health and Welfare" status. Winlink is perfect for this. Individual messages can be uploaded in bulk to a local or remote Winlink server and distributed via an internet link. For others Winlink is available via short wave on HF. It is used by boaters and others on the go.

The message forms are filled out at the shelter and then entered into a preinstalled version of Winlink Express. This program works like common email servers but uses the radio or local internet to distribute the mail. This program allows multiple Health and Welfare messages to be loaded at once and sent hourly or as needed. The recipient is usually not a ham, and goes to a standard internet email address.

The frequency split is enough that Sedan and Winlink can coexist in a properly setup network without issues. In shelters with limited space we just shift Sedan offline briefly and send the Winlink as an hourly burst. Each Winlink message takes only seconds.

17 LEADERSHIP &
DEVELOPING PERFORMANCE EVALUATION EXERCISES

Some people, like me, are very poor candidates for being leaders. We are the least likely ones to be chosen to head up groups or projects; we do not often find ourselves in charge of anything at all. We were not captains of baseball, soccer, or any other kind of teams while growing up....you get the idea. Yet we understand something technical, and we have a passion to improve things in this world.

So this chapter will be (hopefully) a humble attempt to suggest ways to develop **skills, assets and strategies** in a group of volunteers, and to coach that group into performing training exercises that are well beyond what has been done in your area before.

Coaches are truly amazing people. They are able to bring out the best in others around them. They encourage, they cajole, they set higher and higher bars of performance, they criticize, they sometimes bark, and their teams love them for it – because a good coach *serves his or her players,* by helping them achieve something they could never do before.

This is actually an ancient concept, which can be found in the words of Jesus:
"...Whoever wants to be a leader among you must be your servant."
                                                                Matthew 20:26

Serving others is perhaps the best way to be a leader. Serving them by taking the time to teach them ham radio skills and strategies. By helping them build their stations, and solving their problems. This usually requires enormous effort on your part. At the outset, you may not even have the skills yourself (as I did not) and you may have to spend hours and hours testing and trying things as you figure it out yourself. How would you teach others something you haven't figured out yourself? *Stay ahead of them!!*

If you haven't done the FEMA online courses – get them done! If you don't understand the technical aspects, start learning! If your license class isn't way up there, start studying! And then, pass the knowledge along quickly, while it is still fresh and you still remember which parts puzzled you.

Every member of your group is precious. Try not to lose even the difficult ones. Their criticisms may help you spot weaknesses in your approach. Each member will have individual strengths and weaknesses. It is your job to maximize the application of their strengths, and to whittle away at the weaknesses that are holding your group back. Try not to call attention to their weaknesses, but instead find face-saving ways to allow them to shore up the weak parts while still exercising their stronger gifts. Only directly confront when you feel it is absolutely necessary.

TEACHING is one of the best ways to serve others --- if you can nurture the desire to learn. In our case, the more we learned, the *cheaper* we were able to create emergency communications assets locally. That's attractive!

Learning more about antennas allowed us to make many of them for ourselves. Raspberry Pi's are much cheaper than other ways of created a computer controlled station, largely because the operating system (Linux) is free, but there is a learning curve to using Linux and you may have to put in a lot of time learning it, and then passing it on.

We didn't want everyone to have to purchase expensive sound card interface commercial devices, so we learned how to design our own--- and then solder them together. Most of our group had never soldered before! Couldn't tell a resistor from a transistor! Didn't know how to use voltmeters! These basic ham radio skills were gained by hours of patient circuit building. I myself didn't know how to put photos on a web page --- but thanks to **qsl.net** and their suggestions, I learned how to make informative web pages which allowed us to keep educational material right at our fingertips, and also highlight the accomplishments of the group.

When an antenna mount failed on a building rooftop, others who had the necessary talents graciously stepped forward and guided us to fix the broken mount --- but with what we had already learned, we were able to add to that a 2nd station, giving us frequency diversity. We may next add MESH on that rooftop!

Learning how to lay out a printed circuit board and have them made and shipped to us --- made creating new interfaces far, far faster. All the instructional materials that had to be created to teach the group provided good fodder for a book – and the book was even cheaper than zeroxing to hand out more instructional material.

Used 2 meter radios from Ebay, and hours of reading on the Internet of others' experiences helped cut down the costs. You can actually replace blown modular final amplifiers in many 2 meter rigs, only 4 connections to make! (See: https://www.rfparts.com/module.html)

Listening to other mentors is crucial. Dave Welker of the hospital communications group in a nearby town is constantly providing wise advice on how to nudge a group toward better performance. Seek out mentors who have experience that would help your local group grow.

ICS 120a – An Introduction to Exercises	https://training.fema.gov/is/cour seoverview.aspx?code=IS-120.a

It was Dave who set the stage for our most recent amazing success when he challenged me to learn how to write NIMS-compliant exercises. At first, the training seemed to be a waste of time, but it turned out to be quite valuable. Mixing it with the techniques used to train medical doctors with simulation brought about a very ambitious full scale exercise in which hams would be at multiple locations, experience multiple failures (power, antennas, etc) and have scores of pre-written messages to transmit.

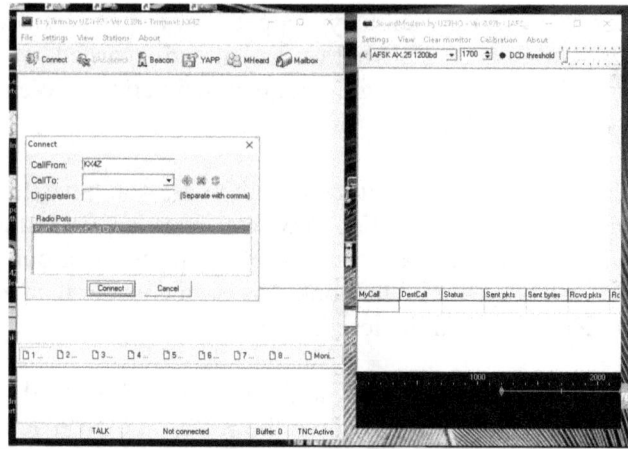

Fig. 17-1 UZ7HO's EasyTerm makes keyboard to keyboard packet QSO's a breeze. Works via an AGWPE connection (tcp/ip port) to Soundmodem

Training for such a complicated full scale exercise took months. In the process we honed our understanding of what systems actually worked the best, tossing out one option in the process and adding a new system (UZ7HO EasyTerm, Fig. 17-1) in midstream (perhaps too confusingly for some).

Developing the paperwork in NIMS-compliant forms took a very long time and still probably wasn't done anywhere close to perfectly correct. But our volunteers understood the basic premise – every hour they would open sealed envelopes and face a new set of handicaps--- and in practice sessions ("tabletop") we worked on each of the skills involved.

Some of our volunteers grasped the importance of portable emergency radios ("go-boxes") and enthusiastically pursued building them. Others took more hand-holding. A bit of woodworking and we had a basic design that made it far easier to carry an entire voice/digital VHF station and modest battery into a facility.

Basic VHF go-box:
- Two 12"x12" roughly 1/2" plywood squares
- 1x4 12" long for sides; 1x6 if more space required
- Use anywhere from 7.5-9 Ahr gel cell for power (use fusing!)
- "Battery Maintainer" devices for keeping battery charged
- Drill one large hole in the top plywood to pass necessary cables from below

- Radio goes on top, easy to reach.
- Souncard interface and battery goes on bottom
- Add some form of SWR meter for output indication if possible.

When you tackle an emergency communications challenge that is so obviously worthy, you feel empowered to start seeking out the necessary resources to succeed at it. Our neighboring Hospital group had done that, concluding that WINLINK digital email was the best emergency communications method, and then seeking funding for the necessary radios and PACTOR and other modems --- and they were rewarded.

In our case, local terrain (100-200 foot high rolling hills) made communications to the western portion of our county quite difficult, yet we needed that to support hurricane shelters. Where to put a repeater node that would cover that portion?

We discovered nearly-abandoned forestry lookout towers that dot the Florida landscape every 15-20 miles, and we learned they were essentially solutions looking for a problem to be solved to be kept from sale and being dismantled --- so we ASKED for support. When you ask honestly and persistently enough, you eventually get answers, but we hit upon the perfect resource on the very first try --- the gentleman in charge of local towers was a very experienced broadcast engineer and ham radio operator who operated several repeaters himself, and recognizing the sincerity of our request (giving him a copy of the 1st version of this book didn't hurt) he gave us full use of exactly what we needed --- a tower in the western portion of our county with a workable antenna and feedline already installed. In one fell swoop we had dramatically improved our digital emergency communications assets!

Fig. 17-2 Forest Grove lookout tower with premounted collinear antenna and heliax cable.....free use given to our group because of our desire to establish emergency packet communications.

(Later testing demonstrated that this one tower can give us communications all the way to a farther tower 40+ miles away, giving us the ability to provide emergency communications to small coastal communities whose normal communications could easily be overwhelmed or damaged by severe storms.)

When our Full Scale Exercise arrived, it seemed like a disaster for the first hour or so --- but then all the hours of training we've done began to kick in as volunteers began to figure out solutions that would actually work from their locations and shelters --- and we ended up with a very impressive success!

Be sure to always put the best face you can on the performance of people you are training. Don't highlight their failures – instead praise them for their successes and then go honestly but gently over the areas that need further work. The Oreo principle – wrap constructive criticism in front and back layers of praise.

Fig. 17-3 Two of our volunteers deployed to a Hurricane Shelter location during our Full Scale Exercise – having already established emergency HF and VHF antennas for the site.

If you will always value the dignity of each of the volunteers who are giving of their time to better serve your community, and put in the work that is truly required to *not waste their time*, but instead to be able to help them reach new heights, you'll develop a winning team.

You can find information on that first full scale exercise that we did here:

Participant Workbook (includes goals, objectives, ICS-205 frequency assignments, planning information	http://qsl.net/nf4rc/ParticipantWorkbook.pdf
Write Up in North Florida Section Newsletter	http://arrl-nfl.org/wp-content/uploads/2014/06/00-QST-NFL-June-2017.pdf
Our After Action Report with Improvement Plans	http://qsl.net/nf4rc/AlachuaCountyARES2017HurricaneTestAfterActionReport.pdf

18 "MESHING IT UP"

MESH ham radio high-speed TCP/IP based microwave networks are the latest rage in Emergency Communications. Ham radio operators want fast, cheap, digital connections that offer many of the advantages that we are all so familiar with in the Internet, available over cable, optical,and even by 4G LTE cell phone radio. 1200 baud AX.25 Packet by comparison is S L O W!

It turns out that MESH communications are quite easily added to the Raspberry Pi **linbpq** systems already described in this book. High speed equipment capable of quite long ranges (when properly applied) is cheaply and readily available in the general public consumer realm and more than one group now offers replacement firmware software that can make the gear "ham-friendly" and in some cases move it to ham-exclusive frequencies as well. But all is not "milk and honey" in MESH-land --- just like any emerging technology, there are pitfalls and problems as well as tremendous opportunities, and the wise emergency communications amateur will do a careful evaluation of how best to use this new technology to add to and complement their pre-existing skills, assets and strategies So this chapter will go over a bit of the background of how to connect to a **linbpq** system.

FIRST....what does "*mesh*" mean? Stations connecting on network can do so in multiple different configurations as shown in the following table:

Topology	Connections
Star (hub and spokes)	Stations connect to central station
Ring	Stations connect to neighbors in a ring
Tree	Hierarchical connections like

	a Christmas tree
Mesh	Full: every station connects to every other station
	Partial: Stations connect possibly to several neighbors

.

In a mesh network, all stations may participate in passing data for the network, and multiple (redundant) pathways are possible. Amateur radio microwave "mesh" networks are generally "partial" because each individual station may not be able to reach all the other stations due to obstructions and distances involved.

Using software, amateur radio mesh networks discover pathways from station to station, assigning figures of merit, so that redundancy develops. If stations disappear, the network can "heal"; if stations are added, the network can grow and improve. The software makes the system self-configuring, so that if a large number of emergency ham radio operators suddenly arrive at a scene, the network can form on the spot. (AX.25, through its nodes broadcasts, does a similar thing.)

The early successful systems apparently took advantage of linux-based LINKSYS model WRT54G (version 1 through 4) wifi home routers, and reflashed their internal software ("firmware") with ham-specific software. This software is available for both LINKSYS and UBIQUITI hardware from Broadband-Hamnet (http://www.broadband-hamnet.org/). Supported hardware is listed here: http://www.broadband-hamnet.org/which-hardware-to-use.html At the time of this writing, their software appears to be at version 3.1.0 in their internal numbering.

Due to a difference of opinion, certain developers from the original group split off and formed a 2nd organization in this same market, AREDN (http://www.aredn.org/) A list of supported platforms for their system can be found here: http://www.aredn.org/content/supported-platform-matrix Their latest software is 3.16.1.1, and apparently the "3" from one firm is compatible with the "3" from the other.

Version 1-4 LINKSYS WRT54G routers are no longer in production, produce about 79 milliwatts of power(19dBm), and can be found on ebay. Current-production mass-produced Ubiquiti products (because of consumer Part 15 regulations) are higher powered, at around 630 milliwatts (28dBm), and are easily available in the $50-$80 range, even on Amazon. The reader will need to do some study to pick units for a beginning, but I was advised to get a NanoStation (not the lower powered -loco version) and an AirRouter HP ("high power," not the lower powered version) and I chose to go with M2 versions, meaning 2.4 GHz. You can of course use any of the available bands; 3 GHz might have been a better choice.

Operation can be done as a Part 15 device on at least portions of the 2.4 and 5 GHz bands, but most hams choose to move to Part 97 where there are fewer power and antenna limitations. Under Part 15 operation,

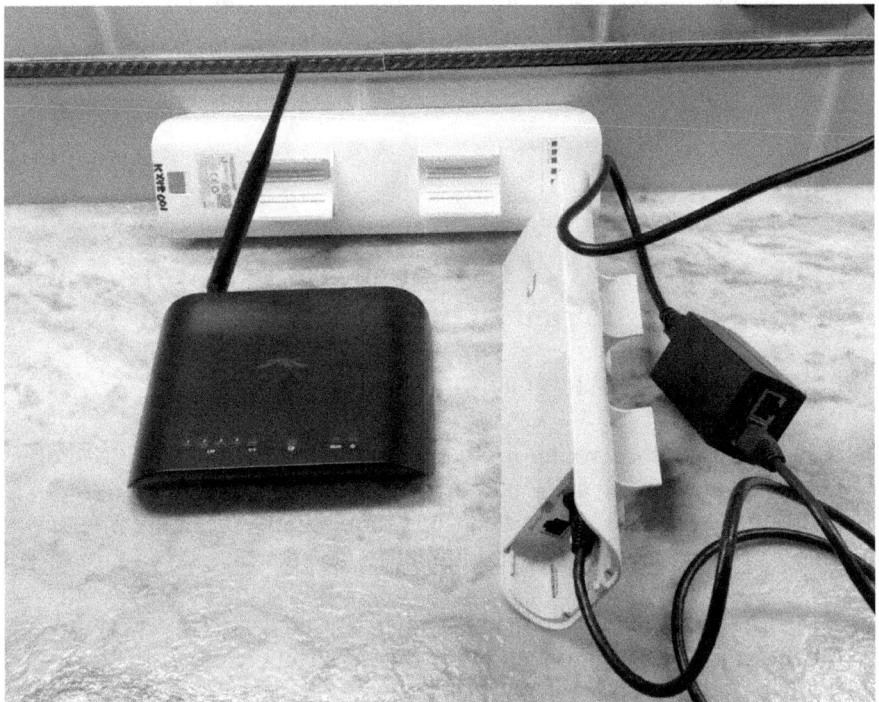

Figure 18-1   Two views of an outdoor-rated Ubiquiti Nanostation (meant to mount on a pole) and also an indoor AirRouter HP, which functions much like any other router.   All gear is powered using Power Over Ethernet (PoE) via two ordinarily unused wires of Ethernet, using the black power supply on the right, which comes with the purchased transceiver.

as you increase antenna gain, you may be required to reduce transmitter output power. Ham radio doesn't have that limitation, but spread-spectrum is still limited to 10W output.

To visualize the overlaps, shared and non-shared frequencies between the ham radio bands and 802.11 Part 15 frequencies, see: https://en.wikipedia.org/wiki/High-speed_multimedia_radio

A LOT OF TCP/IP IN ALL THIS!

In the remainder of the chapter, there are lots of details that involve TCP/IP networking. Readers who are a little less familiar with TCP/IP may need to re-read the earlier chapter on AX.25's information on TCP/IP, or do a bit of investigation into IP numbers, netmasks, etc., for which the Internet provides many resources. (Easy intro here: http://www.cse.wustl.edu/~jain/tutorials/ftp/t_2tcp.pdf *and complicated one here:* http://people.na.infn.it/~garufi/didattica/CorsoAcq/Trasp/Lezione9/TCP_IPFundamentals.pdf)

I. The Nitty-Gritty of Getting a Mesh Network Working and Married to linbpq

TALKING TO THE UBIQUITI DEVICE

Connecting to their router products is easy – they run DHCP and thus automatically can issue an IP number to your PC. Connecting to the NanoStation M2 was a bit harder – had to manually configure my Windows PC to place a static IP number on its ethernet port in the 192.168.1.X range (NOT .20—that's the IP address the NanoStation uses).... Having made the connection, you'll find a webserver on the ip number of the Ubiquiti device to which you're connected.

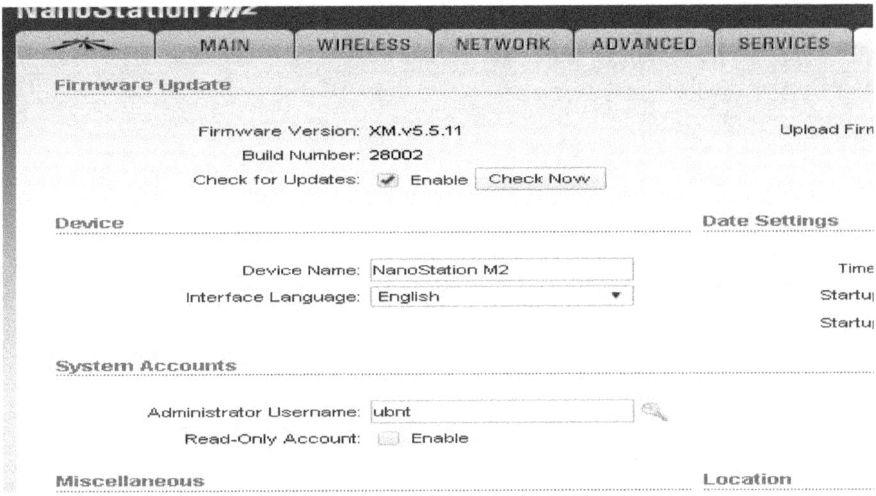

Fig 18.2   Web server interface to a Ubiquiti device.

Through the web server, you can upload new firmware, adjust settings, etc.

CONNECTING TO LINBPQ

This is the important point: It turns out it is easy to marry MESH with linbpq AX.25 technology. My familiarity is only fledgling and only with the Ubiquiti products using AREDN software, but this is all basically TCP/IP and other systems probably work similarly.

Background: **linbpq** is often used with an Internet "tunnel" to allow IP (internet protocol) connections to other **linbpq** nodes that are beyond reach by radio. So it is easy to simply replace the Internet connection (basically a TCP/IP based connection) with two MESH transceivers (also a TCP/IP based connection). In both cases, the AX.25 packets are simply placed inside larger packets appropriate for the TCP/IP network, and passed, then unpackaged at the other end.

In my bpq32.cfg skeleton file, port 2 was obviously slated for this usage.   Filling in the port configuration is relatively simple:

```
PORT
  ID=AXUDP Link
  DRIVER=BPQAXIP
 IPADDR=192.168.1.58
 UDP 10093

  QUALITY=200
  MAXFRAME=4
  FRACK=5000
  RESPTIME=1000
  RETRIES=10
  PACLEN=236
  MINQUAL=150
  UNPROTO=FBB              ; DEFAULT UNPROTO ADDR
  BCALL=KX4Z-7            ; Call for Beacons

  CONFIG
  MAP  KX4Z-2 192.168.1.13 UDP 10093 B
  AUTOADDMAP

  BROADCAST QST
  BROADCAST NODES

  UDP 10093
  # Optional. Enables UDP support, and defines
the port
  # AX.IP listens on. You can specify more than
one
  # UDP line if you need to listen on more than
one port

  MHEARD                    # Optional - opens
a window to display a "Heard List"
  ENDPORT
```

In this case, I'm using UDP (AX.25 will request repeats of any frames that don't come across properly, so the connectionless UDP will become error corrected). UDP is considered optimal for many BPQ systems connecting over a TCP/IP based network. My system is configured to put my MESH hardware on 192.168.1.58, and the station I connect to (KX4Z-2) is configured for their MESH hardware to be 192.168.1.13. AUTOADDMAP will cause any other station heard to be added to this list. If you wish to restrict possible connections, omit

this. More information on the configuration of the port is available from John Wiseman: [76]
http://www.cantab.net/users/john.wiseman/Documents/BPQAXIP%20Configuration.htm

Next connect an ethernet cable (10baseT) from the Raspberry to the RJ45 socket on the Ubiquiti, and as long as the two are set on the same ip network, and the configuration is correct – the Ubiquiti device simply appears as another radio connection for the LINBPQ system!   No need for sound-card interface systems, or TNC....the bpqaxip software handles this!

RADIO FREQUENCY INTEFERENCE
At the time of this writing, all of this is still pretty new to me. With the fragility of the Raspberry Pi (Version 3) I/O ports, I'm concerned about the risk of RFI interference to the tcp/ip connections between raspberry pi and the Ubiquti product. Indeed, using unshielded cables of 3-5 feet in the presence of a 100-watt HF SHARES WINLINK server, I did indeed observe what appeared to be RFI – with the Raspberries losing their DHCP-assigned[xv] address and the Ubiquti's dropping the link as possible symptoms. I would suggest shielded cable and ferrite beads snapped around them as well.

CONSUMER PART 15 OPERATION
I tested the **linbpq** configuration first with the pre-existing consumer configurations of the Ubiquity systems, and was able to easily connect from one linbpq system to another --- the same commands
```
C KX4Z-7 and
MH <port>
```

work just as expected over port 2 (TCP/IP to a Ubiquiti device) as they do over port 7 (2 meters). You may have to take some care (with the Nanostations) to get the Raspberry Pi & Nanostation onto the same ip network and "seeing" each other (try "ping" to test, which sometimes requires root privileges to use).

Having demonstrated that
1. The linbpq Raspberry Pi system can seamlessly communicate

xv DHCP is the service that you probably have used often, which automatically assigns IP address to devices connecting to most consumer routers.

over TCP/IP-based MESH devices

2. The Unbiquiti microwave products can provide a link between **linbpq** nodes even using Part 15 Ubiquiti off-the-shelf firmware. I next proceeded to reflash the Ubiquiti microwave transceivers to use more ham-compatible software. (You probably shouldn't be using ham radio call signs over Part 15 operations....)

SWITCHING UBIQUITI TO HAM-SPECIFIC AREDN SOFTWARE

This takes several steps, and you'd be wise to follow the directions on the AREDN web site. (http://www.aredn.org/content/uploading-firmware-ubiquiti) **NOTE: in all of this, I would recommend that you NOT point microwave devices with high gain antennas at yourself, particularly not at your head.**

Briefly:

- First you have to check to see if your device's firmware version is too recent to work well with the AREDN re-flashing process, using a bit of testing software provided on the AREDN web site. In my case....yep, too recent!
- If too recent, then you "downgrade" your firmware by first downloading the older v. 5.5 of the Ubiquiti firmware (to your computer, helpfully provided by the AREDN web site, using your home Internet connection) and then use the web server on the Ubiquiti device to install that older firmware. Depends on whether you have "XW" or "XM" firmware[xvi]. Don't change any passwords yet (user ubnt, password ubnt)
- Re-run the testing software...get GOOD / GOOD and you're ready to advance. (There are similar warnings and processes discussed and explained on the broadband-hamnet site http://www.broadband-hamnet.org/)
- Since you're starting from a "factory" firmware, go to the AREDN software download page, find the "factory" replacement file for your particular device.
- With the new firmware, the device switches network configurations again, and you have to tell your PC to accept a

---

xvi Note that as of this writing, the AREDN web site has a simple typo and at one point labels the replacement XM software with "XN"...but the alternative is labeled "XW" so it is fairly obvious which to choose.

dynamic IP from your Ubiquiti devic
- Then proceed to the web address given in the AREDN instructions, and put in your callsign and other required configuration information for ham use.

NOCALL-90-57-209

Help [Refresh] [Setup] [Select a theme ▼]

This node is not yet configured.
Go to the setup page and set your node name and password.
Click Save Changes, <u>even if you didn't make any changes</u>, then the node will reboot.

This device can be configured to either permit or prohibit known encrypted traffic on its RF link. It is up to the user to decide which is appropriate based on how it will be used and the license under which it will be operated. These rules vary by country, frequency, and intended use. You are encouraged to read and understand these rules before going

Fig. 18.3   Right after installing ham-based firmware, before configuration. Don't forget to set the slider to the maximum distance you'll ever use – I think it sets a limit on delay times.

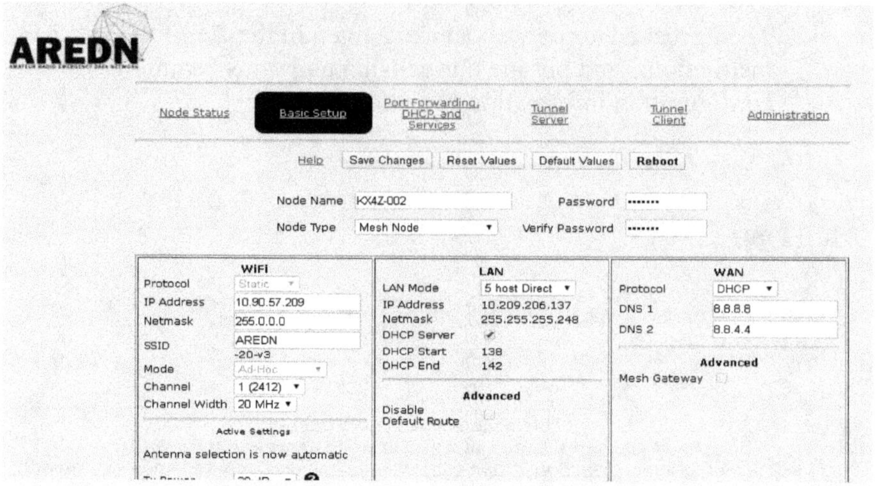

Fig. 18-4 Initial configuration.   Passwords and setting the maximum
distance slider (don't forget that).

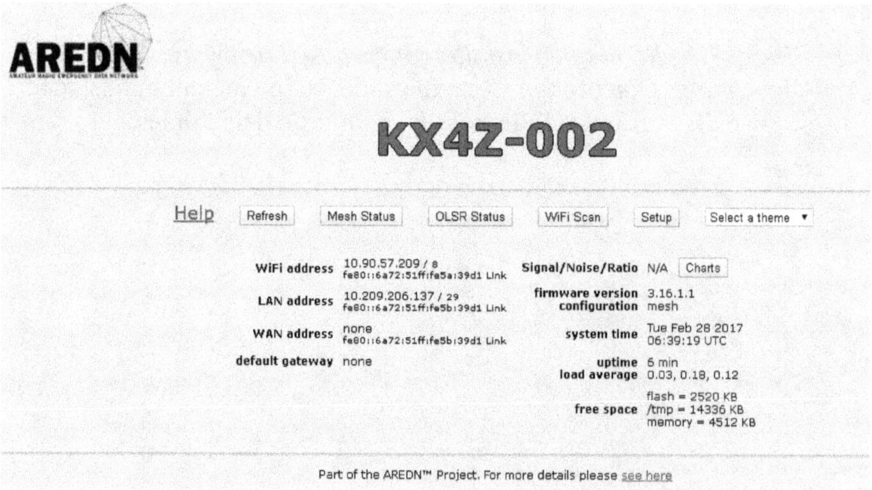

Fig. 18-5   After initial configuration.   It now provides a DHCP server,
and has all the settings you'd expect of a router (port forwarding, etc)
but it also has a lot of ham radio based settings to help you see and
manage what is going on with the MESH network.   If you click on the
WIFISCAN it will show you normal wifi routers in your house (if on
same band).

II. Planning and Implementing a MESH Network

MESH FREQUENCIES AND BANDWIDTHS
There's a problem with WIFI channels and spread-spectrum signals: the typical high speed WIFI signal bandwidth(20-22 MHz) doesn't just occupy *one* WIFI "channel" --- it occupies *several*! And 802.11n uses 40 MHz.

Furthermore, some (not all) of these channels are potentially usable by amateur radio operators, some with other interests....and both amateur & consumer frequency privileges differ by nation! So the reader is advised to carefully consider their usage, bandwidth, and national frequency allocations before proceeding.[77] Even when the frequency can legally be used by amateurs, there are often "band-plans" that recommend some frequencies be reserved for satellite or low-bandwidth applications.

2.4 GHZ BAND:
Relevant ham allocation is 2.390 – 2.450 GHz – meaning that even with a narrow 5MHz bandwidth, the highest usable WIFI channel is at most #8 and probably #7.

WIFI Chan	CenterFreq	USA Part 15?
"-2"	2.397	no
"-1"	2.402	no
1	2.412	yes
2	2.417	yes
3	2.422	yes
4	2.427	yes
5	2.432	yes
6	2.437	yes
7	2.442	yes
8	2.447	yes
9	2.452	yes

10	2.457	yes
11	2.462	yes
12	2.467	yes
13	2.472	yes
14	2.484	yes

It's easy to see if you have a 22 MHz signal bandwidth, and channels only 5 MHz apart, you really can't use all those channels independently! It is common to speak of using Channel 1, 6 or 11. For amateurs, restricting bandwidth to 5 MHz makes it possible to use more of these channels and potentially avoid interfering with satellite operations at and above 2.401 GHz.

An additional wrinkle comes from the ARRL Band Plan[78], as presented on that organization's web page as of this writing: in the 2.4 GHz Band, it is suggested that only the frequencies 2.410-2.450 GHz be used for wideband (> 1 MHz width) signals – meaning that using even modest 5 MHz MESH spread spectrum bandwidth, only channels 2-8 would be recommended for such usage. Channel "-2" falls in an area suggested for "Analog & Digital" of bandwidth 50 kHz to 1 MHz. I'm not an expert on these band plan agreements. The web page indicates that "A band plan refers to a voluntary division of a band to avoid interference between incompatible modes." I would suggest care in using channels/frequencies outside of the "band plan" recommendations. These are somewhat line-of-sight frequencies – you may be able to contact local amateurs with potential conflicts and work out what is the best solution.

3 GHZ BAND:
Relevant ham allocation is 3.300-3.500 GHz – far away from the long-haul 802.11y frequencies.[79] AREDN provides Channels 76 (3.380 GHz) through 99 (3.495 GHz) at 5 MHz spacing. These are all ham radio Part 97 frequencies (in the USA) --- not accessible by the unlicensed public. The ARRL Band Plan suggests the following frequencies for wide-bandwidth signals:
3.310-3.330 No AREDN channels.
3.360-3.400 For 5 MHz bandwidth, channels 76-79
3.460-3.500 For 5 MHz bandwidth, channels 93-99
(Wider bandwidths would require moving inward from those edge channels.)

5 GHZ BAND:

On the 5 GHz band, the USA ham radio allocation is 5.650-5.925 GHz, and the AREDN software is able to offer channels 133(5.665 GHz) - 184 (5.920 GHz), on 5 MHz spacings. ARRL Band Plan suggests frequencies 5.675-5.750 GHz and 5.850-5.925 GHz for wideband signals such as MESH. If (relatively narrow) 5 MHz bandwidths were utilized, these would correspond to channels 136-149, and 171-184. (If wider bandwidth is chosen, the edge channels would need to be moved inward.)

WHY USE BOTH 2-METERS & MICROWAVE?

The reason you would want to be able to marry these two technologies for emergency communications (so that you can move messages easily through either VHF/UHF AX.25 networks, or thorough high speed WIFI-type devices), boils down to the limitation of microwaves compared to VHF --- *far less ability to deal with obstructions.*

MICROWAVE RANGE

While 2 meter radio communications are generally considered "line of sight" there is actually some wiggle room with some diffraction over hills and other obstructions, and some penetration of foliage and obstructions does occur. A highly placed 2 meter repeater can still be utilized when you are inside your house, or in a forest. However, the absorption by these obstructions grows much more significant when moving to 2.4, 3 or 5 GHz, to the point that a common phrase is that microwaves can "go through miles of free space, or 1 tree!" Fresnel zones – a volume of space between two microwave antennas which *should be completely clear of obstructions for optimal passage of radio frequencies* – are widely discussed in MESH circles.

In other words, if you want to build a MESH network for your locale, you really would be happy if your city had no buildings over one story tall, and a huge cliff just outside your city on top of which you could place a microwave MESH repeater with unobstructed direct line-of-sight to every structure in your city --- this would give you good coverage of your city! Flatlands next to mountains seem to do well with MESH, while the rolling hills, and lush forested areas of Florida with many scattered 2-story or higher structures, may have much more difficulty.

Radio Mobile,[80] a free propagation simulation software available on the Internet can be your friend to figure out your likely coverage, either for 2 meters or for microwave MESH systems. I tend to use settings such as the following in my simulations:

Setting	146 Mhz	2397 MHz
Output power	40 watts	0.6 watts Typical Ubiquiti product power
Receiver sensitivity	0.5 microvolts	4 microvolts
Required correct	90.00%	90.00%
Additional dB for "green" color	10 dB	10 dB

The MESH receivers require a much higher input signal voltage for the same signal-to-noise ratio because of their higher bandwidth (which admits more noise).

The theoretical minimum receiver noise power is proportional to the bandwidth in Hz. Taking logarithms and converting to decibels, the noise in dB is proportional to the log of the bandwidth.[81] A 5Mhz wide digital receiver has 500 times the bandwidth of a 10 kHz FM receiver; or about 22 times the voltage (voltage is proportional to the square root of the power) of noise, so the input signal (voltage) needs to be around 20 times that of the narrow band FM receiver for the same signal to noise value.... My suggested Radio Mobile settings may actually be slightly tilted in favor of the WIFI signal.

My simulations for the same location (same antenna height) indicate that a 2 meter AX.25 (40 watts, 4 dB transmission line loss, simple antenna) system will have far more distant coverage and much more solid coverage, than the more line-of-sight microwave system (0.6 watts, 11 dBi antenna, zero transmission line loss because the transmitter is right at the antenna). A key residential 2 meter node station with pretty solid coverage of our city, turns out to have a significantly smaller and much more spotty ("shotgun") coverage prediction when done for the 0.6 watt 2397 MHz signal. .

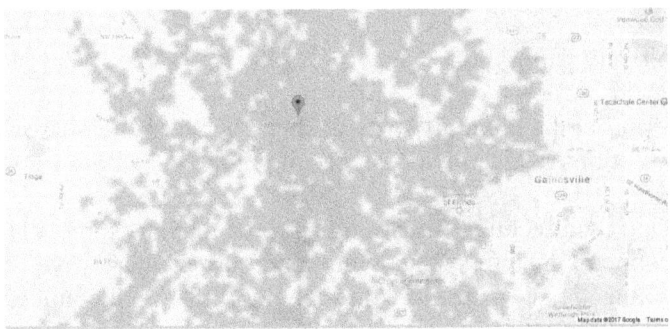

Fig. 18-6 Shotgun pattern MESH coverage from a site in Gainesville Florida that has generally solid 2 meter coverage. Probably due to hills, vegetation and obstructions.

As a result --- the use model for MESH may differ considerably from one geographic location to another. Flat terrain with little vegetation and low-buildings may have great success with a few highly placed MESH nodes and a lot of rooftop ham station nodes. Hill-pocked cities with extensive tree coverage and varying-height structures may find that their MESH covers limited areas and benefits greatly from backup 2 meter AX.25 links.

Disaster sites in open terrain may find significant coverage from a few mesh nodes that are well placed on high points surrounding the area; those in downtown areas with tall buildings may have extremely spotty coverage. Each situation is likely to require a different solution --- but that is the art and science of radio!

FULL DISCLAIMER: I'm quite the beginner at actual integration of MESH networks and **linbpq** networks. It is pretty clear how a simple linear sequential connection, spanning first one distance with 2 meters, and then another with microwaves would work, because there aren't many alternatives. But in a real partial-MESH environment where the MESH network may find several alternative routes between nodes which also have 2 meter transceivers --- it isn't clear to me how **linbpq** will behave-- which "routes" it will choose, whether those via the microwave ports, or those via the 2 meter connections. One can always specify exactly how one wishes connections to be made by placing an exclamation point before the target call sign, e.g.

C 7 !KX4Z-2

to avoid any ambiguity

Furthermore, the great disparity in speed between the microwave and VHF connections leads to two important points:

1. Don't send a huge file through a route that has some VHF connections interspersed with much faster microwave connections --- the slow throughput of the VHF connections means you'll occupy them for a very, very long while;
2. If your network allows a completely MESH based connection to your desired endpoint, keep in mind that your MESH network is a full-fledged TCP/IP high speed network of its own. You might be much better served using the simultaneously available TCP/IP tools that will also work --- TELNET, FTP, HTTP (web browser/web server), pop3 email servers, etc. The broadband-hamnet folks have a page of suggested applications applicable to emergency communications. [82]

Some additional Node Commands that may be helpful as you add MESH networking to a **linbpq** system:[83]

```
---------------Advanced BPQ Node Commands--------------------
LIS <port>     Allows you to monitor any port
UN  <port> CALL   sends UI messages
               Can be used with LIS to hold a multi user chat
N              Show nodes table
N T            shows round trip time & packet counts
```

III. Emergency Communications Bottom Line

Cell phone towers are pretty well constructed. Probably structurally tougher than anything I as an amateur radio operator am likely to install. What exactly is the emergency communications benefit that MESH systems can offer, that would be of use when even the cell towers / Internet / phones are down? That's the central question for any amateur radio emergency technology. I think there are at least two answers:

1. The backup power for the cell towers may be significantly limited, whereas amateur radio emergency-oriented groups may provide very significant backup power systems, thus providing communications when wide-spread power outages have downed cell systems.
2. Weather or other mechanical events that down cell towers (ice, hurricanes, tornadoes) may well also damage amateur systems – but the amateur radio operators may be right there, and able to repair, replace or reposition their antennas more quickly than commercial crews may arrive to fix large numbers of high towers requiring special expertise or equipment to repair

If your group wishes to use MESH microwave technology as part of your emergency communications systems, I would heartily suggest that you keep those two issues firmly in mind, and have backup power and replacement/repair well figured out.

19 MAINTAINING READINESS & GAINING SKILLS & ASSETS

Armies have to continually train for the possibility of war and have to expend time, effort, and material to maintain readiness. Amateur radio volunteers are just that -- amateurs. Maintaining skills at voice is easy -- maintaining skill at digital communications can be much more difficult....but important! A recent QST ARRL Editorial pointed out that analog voice simply isn't going to meet the future needs of emergency communications. The ARRL does a fabulous job of creating frequent opportunities to raise awareness and skill levels at all kinds of emergency communications. Field Day, the Simulated Emergency Test as well as frequent contests give hams lots of chances to work at different skills important in emergency communications.

If volunteers are often having fun with digital QSO's on the HF bands using FLDIGI or similar software, their general digital readiness is likely to remain quite good. EasyTerm packet and WINLINK skills are only somewhat more difficult to acquire and maintain – since the Internet and smartphones have taken over the tasks of providing messaging and instant emails. In my daily life, there aren't many times that I truly need to exercise WINLINK or packet connection skills.

The flip side is that if you really understand digital communications, including WINLINK, it is actually fairly intuitive and not difficult to remember how things work. You have a general idea of roughly how the digital squawks and bleeps are supposed to sound, you remember some strong stations you've used as servers, and you remember roughly what the sequence and timelines are of a contact that is proceeding without a hitch

However, all groups have ebb and flow. People move through different levels of intensity to their hobbies. Family issues can rise and require attention. With older volunteers, health issues can crop up. So your group can dwindle unexpectedly -- best to keep a significant effort at attracting new members and gradually training them up to be able to provide effective communications.

Providing occasional new Technician courses can help. Visiting the local CERT class was incredibly effective for me, as a fast-paced demonstration garnered our group two new Technician students who turned out to be very enthusiastic emergency communications volunteers. Continually encouraging other members to raise their training and licensure level is probably also a good idea. Leaders need to lead -- so make sure you yourself are advancing also!

And always remember what you are doing is important to the safety and lives of many other unknown citizens in your community.

LOOKING BACK over a year of digital EM-COMM training in our local group, it is amazing the transition that is occurring. I don't think that I fully grasped the paucity of knowledge and experience that existed in the local ARES group before we all began this journey. Powerful and highly placed 2-meter voice duplex repeaters had lulled many local hams into almost a CB-type existence -- there was little need for technical expertise, because almost any marginal handitalkie was adequate for accessing the voice repeaters from inside a large region of the city.

Throw in the simple emergency communications question, "What do you do if those repeaters quit working during an emergency?" and a huge amount changed.

Now the importance of the ability to work weaker stations becomes much more apparent and the need for higher antennas, an understanding of feedlines, losses, antenna gain, directional antennas, colinear antennas, higher transmitter power becomes apparent -- and hams gain more knowledge, assets and capabilities. During one training session a local ham suddenly realized the battery pack for his handi-talkie was so degraded that without those repeaters, he couldn't really make any contacts!

Putting up new antennas taught local hams how dipoles are built, how feedlines are connected, what grounding is all about. Most didn't even have an SWR measurement device. Concepts of link budget, free space versus urban building losses were unknown. Many had no expertise with YAGI antennas -- and one new Technician ham had great fun copying a simple design made out of welding rod elements inserted into a PVC boom -- and it worked!

The building of simple audio interface circuitry with plain old soldering techniques brought out the fact that many local hams had never soldered anything together, didn't know how to read schematics, did not know what common components even looked like. Couldn't read resistor color codes, no idea how to position diodes, transistors, LEDs or other polarity-sensitive components! Tremendous growth in technical ability resulted.

There were huge gaps in knowledge of basic computer-usage skills --- copying and pasting text or files, moving about the directories of a disk, downloading and installing new software; dealing with virus scanner software that didn't like unusual new ham radio software --- huge gains in computer technical ability were made in our group and are still ongoing. Actually using software that involved *TCP/IP ports* --- a whole new world for many local emergency communication hams.

Digital packet-based communications made hams think much deeper into signals, emissions, and modes, which many really had only a dim grasp of before.

Beginning to do HF WINLINK communications was a whole new world for many of these hams, who had no experience with Single Side Band, antenna tuners of any type, baluns, "tuning" a final amplifier and checking SWR.

We even had to delve into the area of RF interference when one ham's station was rendered almost "radio blind" by a huge amount of static on the two meter band. A two month effort investigating nearby municipal waste water pumps, power lines and on and on ended with the discovery of a malfunctioning new LED light fixture in the victim's own garage! Enormous learning about other licensed radio services, direction finding, signal strength measurement and radio fox-hunting techniques were the result.

The need to move a much more complicated and more powerful radio station, with self contained batteries and chargers, quickly and easily from home to a service location resulted in hams learning how to put together go-boxes, and deal with 12VDC wiring and fusing needs. Better gear, more reliable, and more technical understanding of electronics and radio.

And now we're adding MESH possibilities to the mix!

Giving people a challenge to become more flexible, better trained, more expert, to move FAR beyond simple mic-button pushing, has been incredibly rewarding for Alachua County emergency ham radio operators. *Your group can do this also!*

APPENDIX ONE: TECHNICAL REFERENCES

Radio Horizon = approx 1.4 * $\sqrt{\text{antenna height}}$
where Radio Horizon is in miles, and antenna height is in feet.

VHF/UHF max range (due only to curvature of earth) is sum of distance to radio horizon from each antenna at the two stations trying to connect.

Max contact due to radio horizon (in miles) for two stations: [84]

Ant. Ht.	10ft.	20ft.	40ft.	80ft	120ft	200ft	300ft
10ft	8	10	13	17	19	24	28
20ft	10	12	15	19	21	26	30
40ft	13	15	18	22	24	29	33
80ft	17	19	22	26	28	33	37
120ft	19	21	24	28	30	35	39
200ft	24	26	29	33	35	40	44
300ft	28	30	33	37	39	44	48

SELECTED FREQUENCY PRIVILEGES
80-10 M BANDS

NOVICE & TECHNICIAN

3.525-3.600	CW	200W OUTPUT
7.025-7.125	CW	200W OUTPUT
21.025-21.200	CW	200W OUTPUT
28.0-28.3	CW, RTTY, DATA; 200W OUTPUT	
28.3-28.5	SSB	200 W OUTPUT

GENERAL CLASS

CW/RTTY/DATA	CW/PHONE/IMAGE
3.525-3.600	3.800-4.000
7.025-7.125	7.125-7.300
10.100-10.150/200 W PEP	
14.025-14.150	14.225-14.350
18.068-18.110	18.110-18.168
21.025-21.200	21.275-21.450
24.890-24.930	24.930-24.990
28.000-28.300	28.300-29.700

ADVANCED CLASS

CW/RTTY/DATA	CW/PHONE/IMAGE
3.525-3.600	3.700-4.000
7.025-7.125	7.125-7.300
10.100-10.150/200 W PEP	
14.025-14.150	14.175-14.350
18.068-18.110	18.110-18.168
21.025-21.200	21.225-21.450
24.890-24.930	24.930-24.990
28.000-28.300	28.300-29.700

EXTRA CLASS

CW/RTTY/DATA	CW/PHONE/IMAGE
3.500-3.600	3.600-4.000
7.000-7.125	7.125-7.300
10.100-10.150/200 W PEP	
14.000-14.150	14.150-14.350
18.068-18.110	18.110-18.168
21.000-21.200	21.200-21.450
24.890-24.930	24.930-24.990
28.000-28.300	28.300-29.700

RESISTOR COLOR CODE

Mnemonic	Color	Value
Better	Black	0
Be	Brown	1
Right	Red	2
Or	Orange	3
Your	Yellow	4
Great	Green	5
Big	Blue	6
Venture	Violet	7
Goes	Gray	8
West	White	9

Transmission Line Losses: Representative Effects of Frequency and SWR: 100 FEET

LINE/FREQ	SWR 1:1	SWR 10:1
RG58/U 10 MHz	1.1 dB	3.71 dB
RG58/U 150 MHz	4.6 dB	9.2 dB
RG8/U 10 MHz	0.6 dB	2.2 dB
RG8/U 150 MHz	2.4 dB	6.1 dB
450 ohm window line (plastic insulated) 10 MHz (dry)	0.2 dB	0.8 dB
450 ohm window line 150 MHz (dry)	0.8 dB	2.9 dB

Transmission Line Voltages[85]

Power	SWR	Highest Voltage	Peak Voltage
50W	1:1	50 VRMS	70 PK
	2:1	71 VRMS	99 PK
	3:1	86 VRMS	119 PK
100 W	1:1	71 VRMS	99 PK
	2:1	100 VRMS	141PK
	3:1	122VRMS	171PK
500W	1:1	158VRMS	210PK
	2:1	224VRMS	313PK
	3:1	274VRMS	383PK

MICROPHONE JACK PINOUT AS VIEWED FROM THE EXTERIOR OF THE TRANSCEIVER

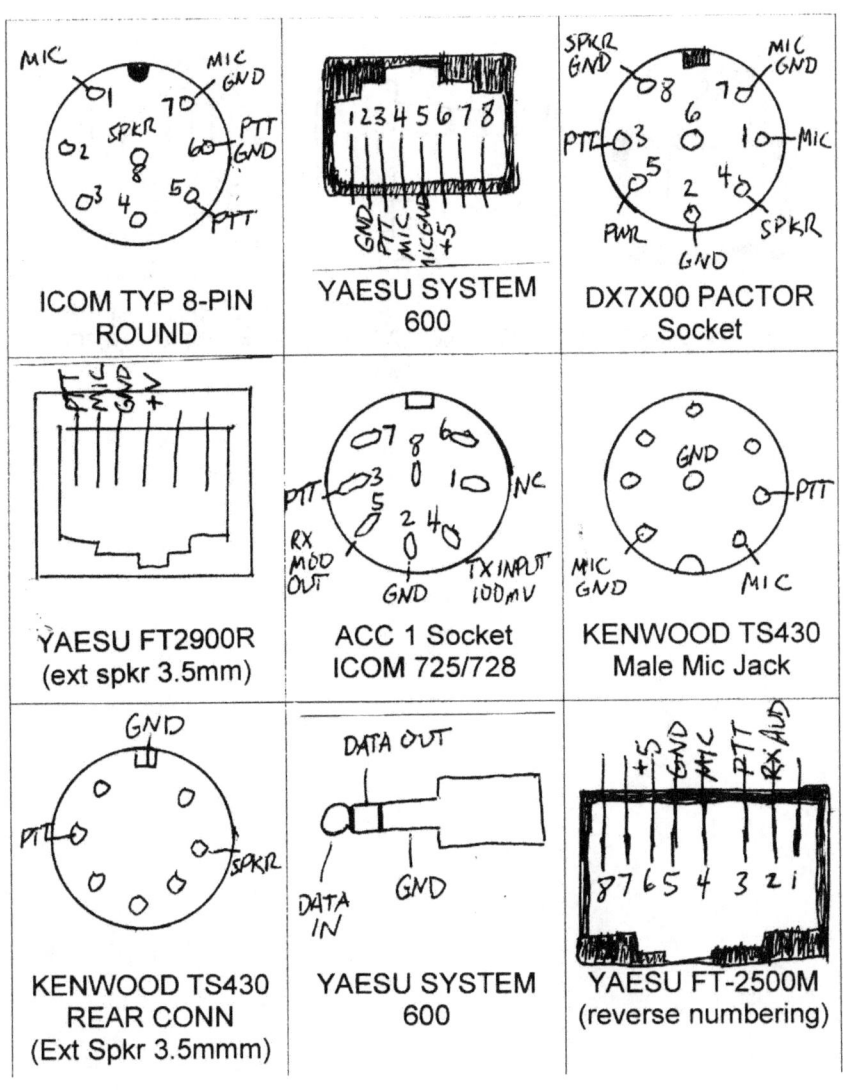

ICOM TYP 8-PIN ROUND	YAESU SYSTEM 600	DX7X00 PACTOR Socket
YAESU FT2900R (ext spkr 3.5mm)	ACC 1 Socket ICOM 725/728	KENWOOD TS430 Male Mic Jack
KENWOOD TS430 REAR CONN (Ext Spkr 3.5mmm)	YAESU SYSTEM 600	YAESU FT-2500M (reverse numbering)

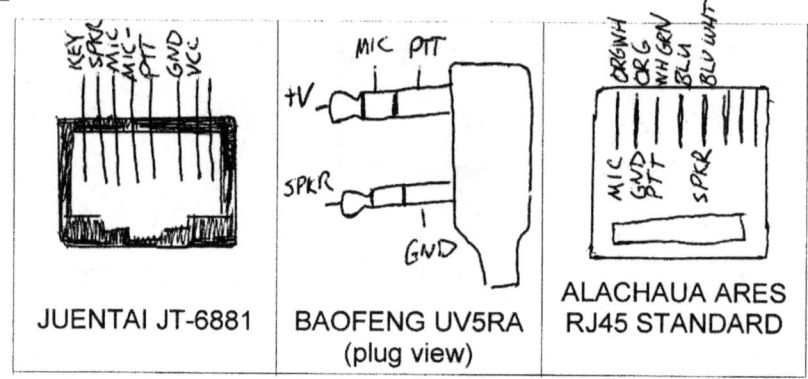

JUENTAI JT-6881 | BAOFENG UV5RA (plug view) | ALACHAUA ARES RJ45 STANDARD

ARRL RADIOGRAM FORMAT

Number   _____   (chosen by originating station)

Precedence   _____

EMERGENCY/P (priority)/ W (welfare)
/ R (routine)   May be preceded by TEST.

HX   _____ [Optional instructions]

Station of Origin   _____

Check   _____

Count of word groups in TEXT, not counting
pro-signs after address and before signature.

Place of Origin   _____

Time Filed _____ Date_____

TO: _____

_____

_____

PHONE NUMBER: _____

EMAIL: _____

("BREAK" / cw BT prior to TEXT)

(BREAK / BT prior to SIGNATURE)

SIG: _____

RECV FROM: _____

Date: _____   Time: _____

SENT TO: _____

Date: _____   Time: _____

This Radio Message Received At:

Amateur Station _____ Phone _____

Name _____ Email _____

Street _____

City, State, Zip _____

REFERENCE:

http://kl7kc.com/radiogramHow2.html

NIFOG REFERENCE:
HTTP://PUBLICSAFETYTOOLS.INFO/NIFOG_INFO/
DOWNLOADS/NIFOG_1_4_J_ROTATED_FOR_VIE
WING.PDF

WEATHER RADIO BROADCASTS – RECEIVE ONLY (WX1-WX7 US & CANADA; WX8-WX9 CANADA MARINE WEATHER)

WX1	WX2	WX3	WX4	WX5	WX6	WX7
162.400	162.425	162.450	162.475	162.500	162.525	162.550

MARINE 21B MARINE 83B

WX8	WX9
161.650	161.775

FRS/GMRS FREQUENCIES

FRS CHANNEL	GMRS CHANNEL	FREQUENCY (MHZ)
1	9	462.5625
2	10	462.5875
3	11	462.6125
4	12	462.6375
5	13	462.6625
6	14	462.6875
7	15	462.7125
8		467.5625
9		467.5875
10		467.6125
11		467.6375
12		467.6625
13		467.6875
14		467.7125
15	1	462.5500
16	2	462.5750
17	3	462.6000
18	4	462.6250
19	5	462.6500
20	6	462.6750
21	7	462.7000
22	8	462.7250

RS-232 Connectors (DB25 and DB9)

"Front" refers to the ends with the pins; "rear" refers to the end with the cable. The following is a view of the pins, looking at the front of the female connector (rear of male):

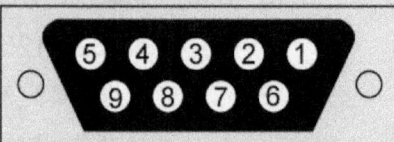

same for DB25, except top pins 13 - 1, bottom 25 - 14 (left to right)

DB9	DB25	Signal
1	8	Carrier Detect
2	3	Receive Data
3	2	Transmit Data*
4	20	Data Terminal Ready*
5	1,7	Ground **
6	6	Data Set Ready
7	4	Request to Send*
8	5	Clear to Send
9	22	Ring Indicator
* An output from the computer to the outside world.		
** On the DB25, 1 is the protective ground, 7 is the signal ground.		

APPENDIX TWO: WINLINK FORMS

WINLINK HTML forms are now a standard part of the continually-updating WINLINK client software, Winlink Express. Forms allow a simple way to send a properly formatted specific type of message which on the screen (and when printed) looks exactly like desired pre-printed form that is useful or important to an agency, hospital or group.

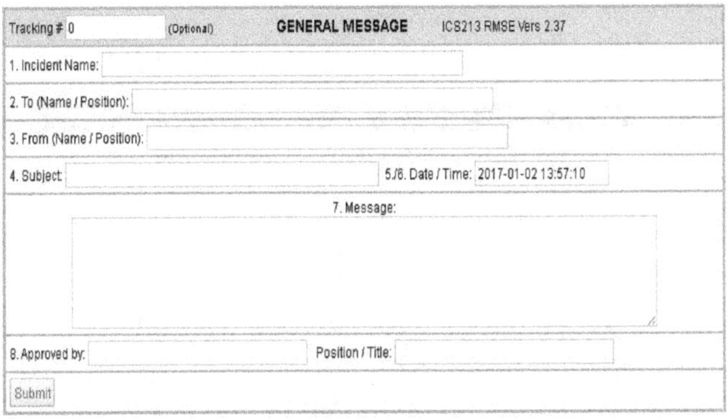

Figure A2-1 *ICS-213 Form. General Message Form*

While normal WINLINK provides you with a "blank page" to type an email message free-form, the html FORMS package replaces the blank screen with a visual representation of a specific form.

The process is simple. From the basic WINLINK email screen, click MESSAGE | New Message, and then on the resulting dialog box, click Select Template, and then follow the list to find your desired template. Once clicked, your default browser will be opened automatically. From there you can fill in the Form as desired, and when finished, click SUBMIT.

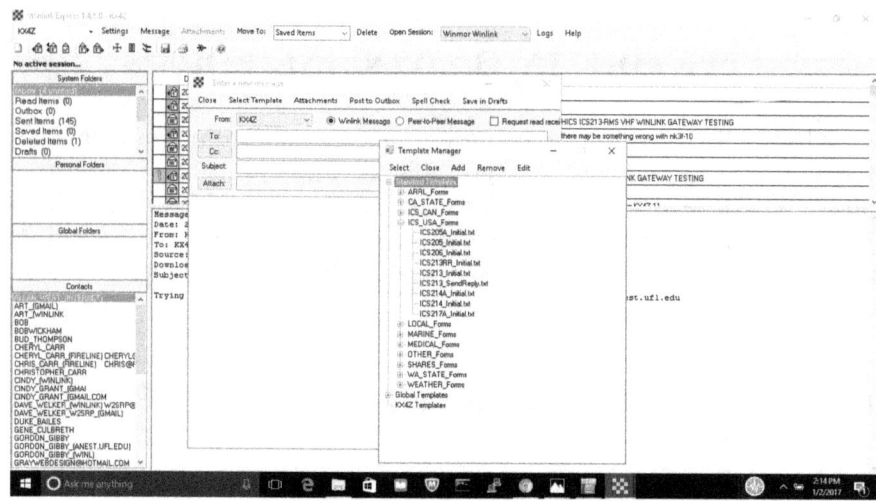

Fig. A2-2   Selecting a Template

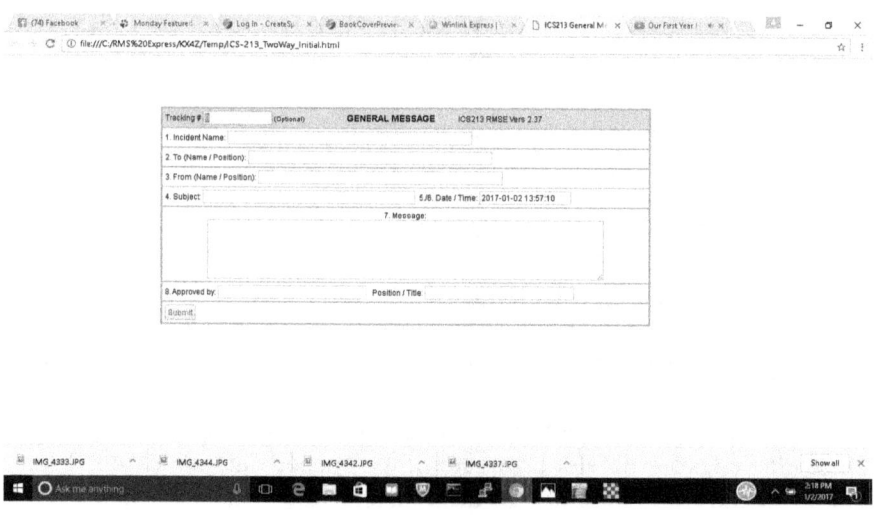

Fig. TR2-3 *Filling in the form inside a standard browser*

You'll be returned to the normal WINLINK e-mail creation dialog box and discover that your message appears in a "text format" but with an XML attachment that contains the information needed to re-create the form at the distant end.

On the receiving end, the WINLINK system will automatically bring up a browser, create the form and populate it with the information sent in the message text.

What happens if the person receiving the email is a non-WINLINK email user?   In that case, the form will not be created, but the text presentation was thoughtfully designed in such a manner to get the necessary information across in an understandable way.

ABOUT THE AUTHOR

Gordon L. Gibby MD has led an eclectic life. Fascinated with electronics at an early age, he was building and even designing amateur radio gear in high school. During his Bachelor's in Electrical Engineering he was a co-op student working with a U.S Government intelligence agency. During his Master's degree he prepared for, and was later accepted, into medical school. Anesthesiology was a natural with all of its electronics, machines, and physiology. An active practice of more than 30 years at a major university teaching hospital put amateur radio on "hold" but did allow him to learn how to fly and become Instrument Rated. Other hobbies include boating, fishing, and target shooting.

Returning back to amateur radio as he neared retirement, he was again fascinated by the possibilities of digital communications to provide effective backup emergency communications in times of severe regional or national distress.

With his incredibly patient wife, Nancy, he has raised three wonderful sons -- all very different-- and has continued to seek ways to serve his fellow man and his Lord and Savior.

REFERENCES

1　EOC Management and Operations, Resource Guide, Dec. 2012, FEMA.　Accessed at: https://www.preparingtexas.org/Resources/documents/TDEM Training/RG_COMPLETE_Dec2012.pdf

2　Congressional Research Service.　The First Responder Network (FirstNet) and Next-Generation Communications for Public Safety: Issues for Congress.　https://fas.org/sgp/crs/homesec/R42543.pdf

3　Communications, A Failure of Initiative.　Final Report of the Select Bipartisan Committee to Investigate the Preparation and Response to Hurricane Katrina, February 15, 2006.　Accessed at: https://www.gpo.gov/fdsys/pkg/CRPT-109hrpt377/pdf/CRPT-109hrpt377.pdf

4　Major General Harold A. Cross,　Adjutant General, Mississippi National Guard.　A Failure of Initiative.　Final Report of the Select Bipartisan Committee to Investigate the Preparation and Response to Hurricane Katrina, February 15, 2006.　p. 174.　Accessed at: https://www.gpo.gov/fdsys/pkg/CRPT-109hrpt377/pdf/CRPT-109hrpt377.pdf

5　See http://www.w0btu.com/wm2u/mt63.html and http://www.arrl.org/mt-63

6　http://www.arrl.org/pactor-iii

7　MFJ-1204 Schematic: http://www.mfjenterprises.com/Downloads/index.php?productid=MFJ-1204P8&filename=1204%20Schematic.pdf&company=mfj ; Manual:　http://www.mfjenterprises.com/Downloads/index.php?productid=MFJ-1204P8&filename=MFJ-1204%20version%203A.pdf&company=mfj

8　http://www.winlink.org/RMSChannels

9　WINLINK History. http://www.winlink.org/content/winlink_history

10 PI-GATE, Emergency Email Gateway, providing Winlink connections using Raspberry Pi client-level software.　Accessed at: http://www.pigate.net/

11 UZ7HO free personal software:　http://uz7.ho.ua/packetradio.htm

12 http://www.winlink.org

REFERENCES

13 Lights Out, Ted Koppel, Crown Publishers, c. 2015, available at: https://www.amazon.com/Lights-Out-Cyberattack-Unprepared-Surviving/dp/055341996X

14 Online FEMA Courses. https://training.fema.gov/is/

15 http://www.arrl.org/online-course-catalog Click on EC-001 Course for further information

16 W4DFU-7, part of the Florida SEDAN Network: http://www.fla-sedan.com/

17 https://weather.com/science/space/news/solar-storm-1859-less-than-day-to-prepare-global-disruption-impact

18 https://www.fema.gov/community-emergency-response-teams

19 There are a plethora of resources for building Slim Jim antennas (which do NOT have any significant gain over a dipole). Here is one: http://www.hamuniverse.com/slimjim.html

20 Comparison of various multiband antenna types: https://www.arrl.org/files/file/Technology/tis/info/pdf/9611073.pdf

21 http://www.tigertronics.com/slusbmain.htm Available from many online retailers.

22 http://www.cantab.net/users/john.wiseman/Documents/InstallingLINBPQ.htm You'll want to do a bit of reading to understand this system.

23 https://www.adafruit.com/product/1475

24 https://portforward.com/

25 Radio Mobile Online, Roger Coude, VE2DBE. http://www.cplus.org/rmw/rmonline.html My experience is that this free online tool gives excellent estimates of VHF coverage.

26 D-STAR, Wikipedia. https://en.wikipedia.org/wiki/D-STAR

27 D-RATS. A Communications Tool for D-STAR And much more.. http://www.d-rats.com/download/doc/contrib/D-RATS_operating_guide_0.3.3.pdf

28 Our Technology -- What is System Fusion? https://systemfusion.yaesu.com/what-is-system-fusion/

29 Broadband-HamNet / HSMM-MESH. http://www.broadband-hamnet.org/

REFERENCES

30 HamWAN. https://www.hamwan.org/

31 Sarasota Digital Network, including a MESH network. http://n4ser.org/sarasota-digital-network-2/

32 UZ7HO soundcard software, both soudmodem97 and easyterm38 (at the time of writing) available at: http://uz7.ho.ua/packetradio.htm

33 MixW. Ham Radio Software from Ukraine. http://mixw.net/

34 UZ7HO. For 1200 baud, select soundmodem97.zip. http://uz7.ho.ua/packetradio.htm

35 Network Wiring Instructions. http://www.cableorganizer.com/telecom-datacom/network-instructions.htm

36 WINLINK via keyboard: http://www.qsl.net/nf4rc/SendingWinlinkWithoutWindows.pdf and also http://n4ser.org/2016/winlink-terminal-program/

37 Introduction to NBEMS, Harry Bloomberg W3YJ, Section Emergency Coordinator, 16 March 2014. Available at: http://www.arrl.org/files/file/On%20the %20Air/Tutorials/Introduction_to_NBEMS_ARRL.pdf

38 Beginner's Guide to FLDIGI, Available at: http://www.w1hkj.com/beginners.html

39 Software by W1HKJ & Associates   http://www.w1hkj.com/

40 www.winlink.org

41 SHARES, https://www.dhs.gov/shares

42 Fact Sheet, https://www.dhs.gov/sites/default/files/publications/SHARES %20Factsheet%20August%202015.pdf

43 SHARES Form 1. https://www.dhs.gov/sites/default/files/publications/SHARES %20Form%201.pdf

44 BUXCOM baluns. https://packetradio.com/catalog/index.php? main_page=index&cPath=11&zenid=d94079c11b1e888b77c7ded01 0587c63

45 AMIDON: http://www.amidoncorp.com/ft-140-61/

46 Balun-How-To   http://www.qsl.net/nf4rc/BalunHowTo.pdf

47 Report of the Commission to Assess the Threat to the United States from Electromagnetic Pulse (EMP) Attack. Volume 1: Executive Report 2004. Figure 2, page 6. Accessed at: http://www.empcommission.org/docs/empc_exec_rpt.pdf

48 Report of the Commission to Assess the Threat to the United States from Electromagnetic Pulse (EMP) Attack. Critical National Infrastructures. Accessed at: http://www.empcommission.org/docs/A2473-EMP_Commission-7MB.pdf

49 All four QST articles are preserved here in PDF format: http://williamesimpson.com/wp-content/uploads/2013/04/QST-Electromagnetic_Pulse_and_the_Radio_Amateur.pdf

50 Faraday Cage, Wikipedia. https://en.wikipedia.org/wiki/Faraday_cage

51 BI-A350 available at http://www.digipart.com/part/BI-A350, and alternate 350-volt gas discharge tube voltage clamp devices available at digikey: : http://www.digikey.com/product-search/en?pv7=3&pv579=4&pv69=80&FV=fff4000a%2Cfff80042&mnonly=0&newproducts=0&ColumnSort=0&page=1&quantity=0&ptm=0&fid=0&pageSize=25

52 MFJ Intelli-tuner design. http://www.mfjenterprises.com/Product.php?productid=mfj-993B

53 Articles on building traps from ordinary coaxial cable: http://www.qsl.net/on7eq/projects/coaxtraps.htm https://www.w8ji.com/traps.htm

54 https://en.wikipedia.org/wiki/Packet_switching

55 https://en.wikipedia.org/wiki/OSI_model

56 https://en.wikipedia.org/wiki/Internet_protocol_suite

57 Address Resolution Protocol: https://en.wikipedia.org/wiki/Address_Resolution_Protocol

58 Ethernet frame: https://en.wikipedia.org/wiki/Ethernet_frame

59 OSI Model: http://www.practicalnetworking.net/series/packet-traveling/osi-model/

60 Function of the TCP layer:

REFERENCES

https://networkengineering.stackexchange.com/questions/27830/wha
t-do-tcp-udp-add-to-raw-ip/27837

61 https://en.wikipedia.org/wiki/AX.25

62 MixW Ham Radio Software from Ukraine. Provides not only
AX.25 packet but also most other popular protocols.
http://mixw.net/

63 UZ7HO soundmodem free soundcard software:
http://uz7.ho.ua/packetradio.htm

64 Direwolf free sound card software:
https://github.com/wb2osz/direwolf

65 https://www.tapr.org/pub_ax25.html

66 TNC-X available at: http://www.tnc-x.com/

67 Raspberry Pi Version 3 available at:
https://www.amazon.com/Raspberry-Pi-RASP-PI-3-Model-
Motherboard/dp/B01CD5VC92/ref=sr_1_3?
s=pc&ie=UTF8&qid=1473881712&sr=1-
3&keywords=Raspberry+Pi+3 and many other vendors as well.

68 I like this particular case better than most others. Available at:
https://www.amazon.com/Raspberry-Model-Protective-Heatsinks-
Clear/dp/B01CDVSBPO/ref=sr_1_1?
ie=UTF8&qid=1473881862&sr=8-
1&keywords=raspberry+pi+3+cases

69 Available at:
https://www.amazon.com/gp/product/B00MWV1TJ6/ref=oh_aui_de
tailpage_o06_s00?ie=UTF8&psc=1 A more expensive alternative
is:
https://www.amazon.com/gp/product/B00FBD3MVA/ref=oh_aui_d
etailpage_o00_s00?ie=UTF8&psc=1

70 Available at:
https://www.amazon.com/gp/product/B00BP5KOPA/ref=oh_aui_det
ailpage_o01_s00?ie=UTF8&psc=1)

71 Terrestrial Amateur Radio Packet Net,
http://tarpn.net/t/packet_radio_networking.html

72 TNC-X: http://tnc-x.com/

REFERENCES

73 ExpressPCB: https://www.expresspcb.com/

74 PiGate: http://www.pigate.net/

75 http://www.hipaajournal.com/hipaa-encryption-requirements/

76 BPQAXIP Configuration: http://www.cantab.net/users/john.wiseman/Documents/BPQAXIP%20Configuration.htm Note the small differences between the Windows references to dll's, which don't apply to Raspberries.

77 Microwave Mesh Channel information: http://www.radio-electronics.com/info/wireless/wi-fi/80211-channels-number-frequencies-bandwidth.php

78 ARRL Band Plan: http://www.arrl.org/band-plan

79 802.11y https://en.wikipedia.org/wiki/IEEE_802.11y-2008

80 Radio Mobile: http://www.cplus.org/rmw/rmonline.html sponsored by TowerCoverage.com

81 http://www.radio-electronics.com/info/rf-technology-design/rf-noise-sensitivity/noise-floor.php

82 TCP/IP Emergency Applications; http://www.broadband-hamnet.org/applications-for-the-mesh.html

83 Advanced node commands: http://www.cantab.net/users/john.wiseman/Documents/NodeCommands.htm   See also: http://www.cantab.net/users/john.wiseman/Documents/Commands.htm

84 Calculator available at: http://www.qsl.net/w4sat/horizon.htm

85 https://www.arrl.org/files/file/Technology/tis/info/pdf/q1106037.pdf